THE
FIRMAMENT
OF
TIME

THE
FIRMAMENT
OF
TIME

LOREN EISELEY

———

NEW YORK ATHENEUM PUBLISHERS

1980

FOREWORD

THE FIRMAMENT OF TIME consists of six lectures which I delivered as Visiting Professor of the Philosophy of Science at the University of Cincinnati in the autumn of 1959. They were the result of a grant made by the John and Mary R. Markle Foundation of New York to the Department of Physiology of the College of Medicine. The purpose of the lectures was to promote among both students and the general public a better understanding of the role of science as its own evolution permeates and controls the thought of men through the centuries. One of the lectures, "How Human Is Man?" was given, in a slightly different version (entitled "The Ethic of the Group") in the Benjamin Franklin Lecture Series at the University of Pennsylvania in 1958. This series will be published in 1960 under the title *Social Control in a Free Society*, edited by Robert Spiller. I owe thanks to the University of Pennsylvania Press for permission to use it in this volume.

This little book does not attempt to pursue all of the intricacies of a full-scale treatment of the evolutionary story, but rather to survey certain aspects of this long drama, as well as to direct the thought of an increasingly urban populace toward nature and the mystery of the human emergence.

I make no apology for my attempt to treat simply of great matters, nor to promote that humane tolerance of mind which is a growing necessity for man's survival.

In appreciation of the cordial hospitality shown me at the University of Cincinnati, I should like to speak with particular affection of Dr. Charles D. Aring and his family, my hosts, who were previously unknown to me, but who became my friends. To them I dedicate this frail remembrance of happy hours spent in their company. Neither would I forget President Walter Langsam, Dean Stanley Dorst, or the many people who, both in and out of the University, made my stay such a pleasant one. Gustav Eckstein trusted me to look at his birds, and busy medical men, such as Dr. Maurice Levine, permitted me to attend their clinics. I was given much more than I gave and for this I am grateful.

LOREN EISELEY

THE UNIVERSITY OF PENNSYLVANIA
FEBRUARY 15, 1960

CONTENTS

vii

The splendours of the firmament of time
May be eclipsed, but are extinguished not;
Like stars to their appointed height they climb,
And death is a low mist which cannot blot
The brightness it may veil. When lofty thought
Lifts a young heart above its mortal lair,
And love and life contend in it, for what
Shall be its earthly doom, the dead live there
And move like winds of light on dark and stormy
air.

PERCY BYSSHE SHELLEY

I

HOW THE WORLD BECAME NATURAL

That then this Beginning WAS, *is a matter of faith, and so infallible.* WHEN *it was, is matter of* REASON, *and therefore various and perplex'd.*

JOHN DONNE

M A N is at heart a romantic. He believes in thunder, the destruction of worlds, the voice out of the whirlwind. Perhaps the fact that he himself is now in possession of powers wrenched from the atom's heart has enhanced the appeal of violence in natural events. The human generations are short-lived. We have difficulty in visualizing the age-long processes involved in the upheaval of mountain systems, the advance of continental glaciations or the creation of life. In fact, scarcely two hundred years have passed since a few wary pioneers began to suspect that the earth might be older than the 4004 years B.C. assigned to it by the theologians. At all events, the sale of Velikovsky's *Worlds in Collision* a few years ago was a formidable indication that after the passage of two centuries of scientific endeavor, man in the mass was still enormously susceptible to the appeal of cataclysmic events, however badly sustained from the scientific point of view. It introduced to our modern generation, bored long since with the endless small accretions of scientific truth, the violence and catastrophism in world events which had so impressed our forefathers.

Man has always had two ways of looking at nature, and these two divergent approaches to the world can be observed among modern primitive peoples, as well as being traceable far into the primitive past. Man has a belief in seen and unseen nature. He is both pragmatist and mystic. He has been so from the beginning, and it may well be that the quality of his inquiring and perceptive intellect will cause him to remain so till the end.

Primitive man, grossly superstitious though he may be, is also scientist and technologist. He makes tools based upon his empirical observation of the simple forces around him. Man would have vanished long ago if he had been content to exist in the wilderness of his own dreams. Instead he compromised. He accepted a world of reality, a natural, everyday, observable world in which he existed, and whose forces he utilized in order to survive. The other aspect of his mind, the mystical part seeking answers to final questions, clothed this visible world in a shimmering haze of magic. Unseen spirits moved in the wood. Today in our sophistication we smile, but we are not satisfied with the appearances of the phenomenal world around us. We wish to pierce beneath to ask the question, "Why does the universe exist?" We have learned a great deal about secondary causes, about the *how* of things. The why, however, eludes us, and as long as this is the case, we will have a yearning for the marvelous, the explosive event in history. Indeed, so restless is man's intellect that were he to penetrate to the secret of the universe tomorrow, the likelihood is that he would grow bored on the day after.

A scientist writing around the turn of this century remarked that all of the past generations of men have lived and died in a world of illusions. The unconscious irony in his observation consists in the fact that this man assumed the progress of science to have been so great that a clear vision of the world without illusion was, by his own time, possible. It is needless to add that he wrote before Einstein, before the spread of Freud's doctrines, at a time when Mendel was just about to be redis-covered, and before advances in the study of radio-activity had made their impact—of both illumina-tion and confusion—upon this century.

Certainly science has moved forward. But when science progresses, it often opens vaster mysteries to our gaze. Moreover, science frequently discovers that it must abandon or modify what it once be-lieved. Sometimes it ends by accepting what it has previously scorned. The simplistic idea that science marches undeviatingly down an ever broadening highway can scarcely be sustained by the historian of ideas. As in other human affairs, there may be prejudice, rigidity, timid evasion and sometimes inability to reorient oneself rapidly to drastic changes in world view.

The student of scientific history soon learns that

a given way of looking at things, a kind of unconscious conformity which exists even in a free society, may prevent a new contribution from being
followed up, or its implications from being fully
grasped. The work of Gregor Mendel, founder of
modern genetics, suffered such a fate. Darwin's
forerunners endured similar neglect. Semmelweis,
the discoverer of the cause of childbed fever, was
atrociously abused by his medical colleagues. To
rest uneasy consciences, we sometimes ascribe such
examples of intolerant behavior to religious prejudice—as though there had been a clean break, with
scientists all arrayed under the white banner of
truth while the forces of obscurantism parade under the black flag of prejudice.

The truth is better, if less appetizing. Like other
members of the human race, scientists are capable
of prejudice. They have occasionally persecuted
other scientists, and they have not always been able
to see that an old theory, given a hairsbreadth
twist, might open an entirely new vista to the human reason.

I say this not to defame the profession of learning
but to urge the extension of education in scientific
history. The study leads both to a better understanding of the process of discovery and to that
kind of humbling and contrite wisdom which
comes from a long knowledge of human folly in a

field supposedly devoid of it. The man who learns
how difficult it is to step outside the intellectual
climate of his or any age has taken the first step on
the road to emancipation, to world citizenship of a
high order.

He has learned something of the forces which
play upon the supposedly dispassionate mind of the
scientist; he has learned how difficult it is to see dif-
ferently from other men, even when that difference
may be incalculably important. It is a study which
should bring into the laboratory and the classroom
not only greater tolerance for the ideas of others
but a clearer realization that even the scientific atmosphere evolves and changes with the society of
which it is a part. When the student has become
consciously aware of this, he is in a better position
to see farther and more dispassionately in the guidance of his own research. A not unimportant by-
product of such an awareness may be an extension
of his own horizon as a human being.

I have sought to emphasize this point in my beginning discussion because several of my topics will
be involved with the intellectual climate of the
past. In conclusion I hope to venture some comment upon the world we now call "natural," as if,
in some manner, we had tamed it sufficiently to in-
clude it under the category of "known and explored," as if it had little in the way of surprises yet

in store for us. I shall want to look at this natural world both from the empirical point of view and from one which also takes into account that sense of awe and marvel which is part of man's primitive heritage, and without which man would not be man.

For many of us the Biblical bush still burns, and there is a deep mystery in the heart of a simple seed. If I seem for a time to be telling the story of how man came under the domain of law, how he reluctantly gave up his dreams and found his own footsteps wandering backward until on some far hillside they were transmuted into the footprints of a beast, it is only that we may assess more clearly that strange world into which we have been born —we, compounded of dust, and the light of a star.

Our first effort will be devoted to an examination of that universe which, in the unconsciously prophetic words of Sir Thomas Browne, "God seldom alters or perverts, but like an Excellent Artist, hath so contrived his work, that with the self-same instrument, without a new creation, he may effect his obscurest designs." When the great physician uttered those words in 1635 he was not thinking of evolution, but, as we shall see, he spoke like a blind oracle. The "self-same instrument" effecting design without creation would go unnoticed for two hundred years.

II

It is a moot question in history what brought into existence man's earliest conception of natural law. Certainly what we might call the regularities of nature—the round of the seasons, the passage of day and night—man must have been aware of since the time he began to think at all. There is, however, a difference between this background of observation and the development of the idea that the entire universe lay under some divine system of ordinances which were unalterable.

Whitehead has contended that the medieval insistence on the rationality of God—something which had its origin in the union of Judeo-Greek philosophy—lies at the root of science. On the other hand, it must be borne in mind that the immediately prescientific era was fascinated by monsters, signs in the heavens—miraculous events which a modern man would regard as unlawful and outside the normal course of nature.

It has also been suggested that the rise of centralized royal authority and the extension of its power in the capitalist state in some manner contributed to th. conception of far-ranging law wielded by the Divinity. Like most purely economic explanations of intellectual events, this is doubtless a simplification of a complicated shift in intellectual emphasis. At all events, in the sixteenth and seventeenth centuries, law, natural law, the undeviating law of God, had taken precedence in intellectual circles over the world of the miraculous. There was a rising interest in the Second Book of Revelation; that is, nature. It was assumed that the two books, separately examined, would bear each other out—that the world could be read as though it were part of the Great Book. Since the God of the Old Testament was a God of wrath, it is not surprising that there lingered in the western mind a taste for the violent interpretation of geological events. Across eighteenth-century Europe lay the fallen, transported boulders of what seemed the visible evidence of some vast deluge. In the story of those stones, man's naïve faith in visible catastrophe is countered by the magnificent violence hidden in a raindrop.

III

Time and raindrops! It took enormous effort to discover the potentialities of both those forces. It took centuries before the faint trickling from cottage eaves and gutters caught the ear of some inquiring scholar. Men who could visualize readily the horrors of a universal Flood were deaf to the roar of the invisible Niagara falling into the rain barrel outside their window. They could not hear it because they lived in a time span so short that the only way geologic change could be effected was by the convulsions of earthquakes, or the forty torrential days and nights that brought the Biblical Deluge.

The world of medieval thought was deeply centered upon itself and upon the traditional myths of Christianity. In spite of sectarian clashes, Christians of the prescientific era saw the earth essentially as the platform of a divine but short-lived drama—a drama so brief that there was little reason to study the stage properties. The full interest centered upon man—his supernatural origins, the drama of his Fall from the deathless Garden, the coming of his Redeemer, and the day of his Judgment.

Outside space was the Empyrean realm beyond time and blemish. Inside were corruption and a falling away from grace which were the consequence of man's sin. The atmosphere was not one to encourage scientific exploration. Men were busied about their souls, not about far voyages either in space or in time. They were contented with the European scene; they were devout and centered inward. It was indeed a centripetally directed society on an earth which itself lay at the center of the universe. Sinful though man had proved to be, he was of enormous importance to himself. The eye of God was constantly and undividedly upon him. The Devil, passing to and fro upon Earth, contended for his soul. If man was not in all ways comfortable, he was at least valuable to divinities, and good and evil strove for the possession of his immortal being.

Then someone found a shell embedded in rock on a mountain top; someone saw the birth of a new star in the inviolable Empyrean heavens, someone watched a little patch of soil carried by a stream into the valley. Another saw a forest buried under ancient clays and wondered. Some heretical idler observed a fish in stone. All these things had doubtless been seen many times before, but human interests were changing. The great voyages that were to open up the physical world had begun. The first telescope was trained upon a star. The first crude microscope was turned upon a drop of ditch water. Because of these small buried events, a world would eventually die, only to be replaced by another—the world in which we now exist.

In the early days of scientific exploration of the universe, the divergence between the world of belief as represented by Biblical tradition and the world of science was not anticipated. It was generally assumed that the investigation of the physical universe would simply reveal more of God's ways toward the care of man and reaffirm Biblical truths. The title of John Ray's well-known book, *The Wisdom of God Manifested in the Works of the Creation,* which first appeared in 1691, suggests this outlook. It did not prevent sharp observations from being made, nor a growing and persistent wonder about fossils, which were then called "formed stones."

Erosion was beginning to be faintly glimpsed as a power at work in nature. "That the height of the mountains doth continually diminish," muses John Ray, "is very likely." Our knowledge of time would have to be greatly altered before anyone would inquire whether persistent forces might be at work which would prevent the total denudation of the land and its eventual disappearance into the engulfing sea.

The change which was to pass over human thinking, however, began in the skies. This astronomical thinking has both its conservative and liberal side.

In the history of science, as in other history, there is rarely a specific place to begin a new story. For our purposes I will select just two events: the discovery of the speed of light, and Newton's formulation of the laws of gravitation. If we were to go farther back in time, we would, of course, have to treat of the opening out of space and the intellectual revolution introduced by the discovery that the earth was not the center of the solar system. Nevertheless, it will suffice for our purposes if we recall that it was a seventeenth-century astronomer, Olaus Roemer, who first deduced the speed of light in 1675. In doing so he opened a doorway upon limit-less vistas in space and time. It had previously been supposed that the movement of light was instanta-neous. When Roemer discovered a slight lag in the reappearance of one of Jupiter's satellite moons after an eclipse, it became possible, through the calculations of later workers, to estimate the speed of light at 186,000 miles a second. The way now lay open to the light-year—to astronomical time the magnitude of which lay beyond human comprehension. For a period, in fact, time would remain the possession of the astronomers and haunt only the trackless abysses of space. It would not dis-turb the man in the street.

For that matter, astronomy still reflected much of its ancient attitude toward Empyrean space as inviolable and unchangeable. Newton, a deeply re-ligious man, was of this persuasion. Kepler had written earlier that the celestial machine is not something like a divine organism, but rather "some-thing like a clockwork in which a single weight drives all the gears." Newton, with his formulation of the laws of gravity, had supplied the single weight. God had been the Creator of the machine, but it could run without his interference. At the most, only an occasional interposition of his power would be needed to set the clock right. In contrast to earlier periods, knowledge of natural forces led to less need for divine intervention in earthly affairs. Newton, however, remained devout in a way that many of his followers of the eighteenth century did not.

The growing interest in mechanics throughout the eighteenth century, the passionate fondness for mechanical devices of all sorts, led to an enthusiastic interest in the *Machina Coelestis*. Miracle was in the process of disappearance save at the moment of creation. The world of science was growing increasingly skeptical as its knowledge increased. Signs in the heavens, wonders in the animal world, were decreasing. The machine reigned. God, who had set the clocks to ticking, was now an anomaly in his own universe. The question of celestial and earthly origins, which Newton had abjured, emerged first in astronomy. It would be Immanuel Kant and the French skeptic Laplace who would introduce cosmic evolution, and who would extend backward into time the laws which Newton had extended across space. The wheels and cogs of the celestial machine for the first time would be pursued backward until they dissolved in spinning vapor. By the midpoint of the century, time was clearly seen as necessary to the development of the cosmos. Thomas Wright had identified as a galactic universe the Milky Way, of which our sun is a minor inhabitant.

The philosopher Kant, drawing inferences of galaxial rotation from Edmund Halley's detection of star movement in 1717, proposed the nebular hypothesis of star and planetary origin in clouds of rotating gas. Kant, at last, was seeking to derive the complex from the simple—"the simplest," as he wrote, "that can succeed the Void."

At the close of the eighteenth century, Laplace, in his *Treatise on Celestial Mechanics,* greatly elaborated the nebular hypothesis. With the concept of historical change and development applied to the heavens, the notion of creation by divine fiat in a universe of short duration began to pass. The way was opening for geologists to pick up the story of the molten earth and carry its development forward into time. Without anyone's knowing the precise way in which the change had been effected, the intellectual climate was altering.

By the 1750's cosmic evolution was openly discussed; geological change, timidly; the evolution of life, in subdued and sporadic whispers. As the idea reached our planet, so to speak, it was greeted with less enthusiasm. It aroused curiosity among the masses but seemed to threaten entrenched religious institutions. It is perhaps not without significance that the chief proponent of cosmic evolutionism arose in radical and free-thinking France, as did two of the first great biological evolutionists.

The devious threads of communication which eventually combined all these ideas of development by physical forces, rather than accepting the theory of creation at the direct hands of a Master Mechanic, are now difficult to trace. Eighteenth-century scientists corresponded, where today they would send a paper or a note to a professional journal. Or they paid visits to each other, or chatted at courts and salons. Moreover, the footnoting of the sources of ideas had not become the traditional practice that it is today.

Whatever the methods used, ideas of development, change, what we would call "historicity," appear in several distinct fields with surprising rapidity, if not simultaneously. In later chapters we shall consider the penetration of this same idea into the life sciences and how it came to extend itself to man. Here, however, I wish merely to examine the way in which Newton's conception of the cosmic machine, the celestial engine, came to extend itself into geology. So far as the life sciences are concerned, we shall see, later on, that the whole idea was both advantageous and, paradoxically, retarding. The flow of ideas from one field into another often takes curious and ambivalent paths. It was so in the case of James Hutton, the founder of historical geology.

IV

THE scientific life of James Hutton extended over the last half of the eighteenth century. What Newton achieved and emphasized in astronomy and mathematics, Hutton accomplished in geology. It would be useless here to pursue all of the faint hints and intuitions about geological matters which preceded Hutton's work. They exist, but they do not lessen the fact that it was James Hutton of Edinburgh who, in diligent application of Newton's principles of experimental inquiry and observation, passed from the astronomer's conception of the self-correcting machine of the heavens to the idea that the earth itself constituted a machine which eternally reconstituted and renewed itself.

It has been pointed out that Hutton's doctoral dissertation was on the circulation of the blood in the Microcosm, that is, in man. It is an old idea in western thought, which persisted with unabated force into the eighteenth century, that man the microcosm reproduces in miniature, or is directly influenced by, the events of the Macrocosm, that is, the outside world, the universe. This idea lies at the root of astrology, and persists in a disguised form into modern times. It has been contended that Hutton, as a medical man, applied this idea to the earth, treating it as a living organism with circulation, metabolism, and other correspondences to the organic world. It has been termed Hutton's secret—a secret which happened to yield, in the case of geology, some remarkable insights, because it placed emphasis upon the dynamic qualities of the earth's crust—in short, upon the phenomena of decay and renewal.

It can be maintained, however, that just as Newton and his successors placed emphasis upon the giant celestial machine of the heavens—self-balancing and self-maintaining, set rolling by the hand of the Master Craftsman, God—so Hutton, influenced by this widespread conception, was the first to apply it to another seemingly self-renovating engine under divine care, the earth. Though Hutton, in later years, was not to escape the charge of heresy, the existing documents clearly suggest that he was less a cosmic evolutionist like Laplace than he was a true Newtonian, in that he abjured, or at least evaded, the question of the earth's origins. Rather, he dealt with the planet as a completed mechanism—whether we regard that mechanism as organic in essence or mechanical. He had accepted and read in the rocks, as the astronomers had begun to read in the skies, the message of time. Indeed he states forthrightly of the earth "that we find no vestige of a beginning, no prospect of an end."

Which, then, of these two views of Hutton's

achievement is correct? Is Hutton's a machine analogy or an organismic one? The machine analogy,
at any rate, bulks large in the interpretation of
eighteenth-century thought and descends into our
own day. It is only by the hook of the analogy, by
the root metaphor, as one philosopher has termed
it, that science succeeds in extending its domain.

Occasionally, if not frequently, the analogy is
false. Yet so potent is its effect upon a whole generation of scientific thinking that it may lie buried in
the lowest stratum of accepted thought, or color
unconsciously the thinking of entire generations.
While proceeding with what is called "empirical
research" and "experiment," the scientist will almost inevitably fit such experiments into an existing comprehensive framework, an integrative for-
mula, until such time as that principle gives way to
another. Let us see, in this connection, what ideas
Hutton introduced into his examination of a previously neglected subject: the nature of the habit-
able earth.

V

WE HAVE earlier spoken of man's individual life
views as colored and influenced by what he can
perceive within that lifetime—what, in other
words, he can personally observe. The individual

is loath to accept explanations of phenomena which
come about as the result of forces exceeding the
range of his own life span. If it is some type of natural landmark placed before his day, he is apt,
rather than consider the effect of the accumulation
of small events, to turn to myths incorporating out-
right violence on a gigantic scale. This, as we have
already observed, is the first natural reaction of
many laymen unacquainted with the history of geology today. In the seventeenth and eighteenth centuries science attempted its first groping entrance
into the vast domain of time. It is not surprising,
therefore, that what is now confined to the naïve
and scientifically uneducated should have affected
the reasoning even of scholars.

The Christian world accepted a surprisingly
short time scale of a few thousand years. The calculations of such men as Bishop Ussher, based upon
genealogical charts and other stray Biblical sources,
were not an integral part of the Bible, but through
long association with the volume they had become
so. Moreover, the story of Creation, Eden, and the
succeeding Flood all imply a world controlled and
brought into being by direct supernatural methods
which seemed to be devoted solely to the human
drama. These had been the cherished beliefs of
Christendom for over a thousand years. They were
graven deep in the religious consciousness of scientist and layman alike. The Book of Nature sought

by deists and religious liberals was an embodiment of divine reason and would not contradict the other great source of direct revelation—the Bible.

The result was that when Hutton, again under the influence of Newton's mathematical analysis of continuity, postulated the integration of small events to produce great cumulative ones in geology, he differed sharply from his associates. Doubtless these ideas contributed strongly to the charges of atheism which were hurled at him. Certainly it was not long before his views were utterly at odds with that school of thought known as catastrophism, which was destined to obscure his work for a whole generation.

This school of violence is the very antithesis of the Huttonian approach through time, raindrops and aerial erosion.

The catastrophist believed the glacial boulders scattered far from their point of origin to have been rolled and tossed in the turbulence of some giant deluge like the Noachian Flood—visible evidence of wild powers loosed upon the planet at sporadic intervals. Mountain chains were the product of similar violence. Breaks in the geological record, discontinuities, in time even abrupt faunal changes, were all assumed to be the devastating result of world-wide disturbances.

As geological knowledge of the earth's history

increased shortly after Hutton's time, this theory, or modifications of it, became ascendant in geological circles. It had about it a certain awe-inspiring Old Testament grandeur. It predicated vast, unknown and perhaps supernatural forces at work. Each cataclysm shut one such geological period off almost totally from another. It was the one great Biblical event multiplied by a chain of such events extending backward into the past. A series of shut doors concealed one age from another.

The only continuity, so far as the living world was concerned, lay in an abstract plan, a Platonic ideal in the mind of God, which caused the beings of one age to have an organic phyletic structure related, though only immaterially and with modifications, to the creatures of another.

Catastrophism is one of the prime examples of a scientific world view in transition. Its mysterious geological upheavals and re-creations of life could be paced fast or slow according to the Biblical days of creation as figuratively expressed in the Book of Genesis, or as the tolerance of the individual might incline. Its succession of convulsive movements of the earth's crust accounted for the more dramatic aspects of the European countryside without introducing those limitless and invisibly moving landscapes which seemed, to many Englishmen in the early years of the nineteenth century, to be part of the dreadful culmination of heretical thought as it had afflicted France.

Political and religious considerations aside, however, catastrophism has an appeal of its own, even into our own day. No one likes to watch, listlessly, an hour hand go around the clock. We want the cuckoo bird to erupt violently at intervals from his little box, or a gong to strike. This catastrophism provided. Its time scale was scored and punctuated by violence.

Hutton, on the other hand, presents us with a quite different system. Instead of beginning with ancient catastrophes postulated upon giant tidal waves, he states with the utmost sobriety that "we are to examine the constructions of the present earth, in order to understand the natural operations of times past. The earth," he says, "like the body of an animal, is wasted at the same time that it is repaired. It has a state of growth and augmentation; it has another state, which is that of diminution and decay. This world is thus destroyed in one part, but it is renewed in another." Across Hutton's pages pass a series of small natural operations that over long time periods erode mountains, create valleys, and that, if mountain-building processes did not counteract their effect, would bring whole continents down to sea level.

He saw the bit of soil carried away by a mountain

brook or a spring freshet lodge in and nourish a lower valley; he saw the wind endlessly polishing and eroding stones on the high flanks of the world. He saw, with the marvelous all-seeing eye of Shakespeare, that "water-drops have worn the stones of Troy and blind oblivion swallowed cities up." He knew about the constant passage of water from sea to land and back again. If a leaf fell he knew where it was bound, and multiplied it mentally by ten thousand leaves in ten thousand, thousand autumns. One has the feeling that he sensed, on his remote Scottish farm, when frost split a stone on a winter night. Or when one boulder, poised precariously on a far mountain side, fell after a thousand years. For him and him alone, the water dripping from the cottagers' eaves had become Niagaras falling through unplumbed millennia. "Nature," he wrote simply, "lives in motion." Every particle in the world was hurrying somewhere, or was so destined in the long traverse of time.

In his observation that land was being created while land was being worn away, that there was continental elevation as well as denudation, Hutton shows a great grasp of the earth's interior powers. Though it was impossible for him to be totally correct in small details, he was almost alone in his recognition of the geostrophic cycle.

James Hutton had come upon the secret of the

relatively perpetual youth of the planet. Although Hutton was primarily a physical geologist who published little upon fossils, he had, in actuality, set the scene in which, a half-century farther on, the rise of the vertebrates might be better grasped. In fact, he had provided the physical setting for an evolutionary process as lengthy in its implications as his own eroding hills.

Hutton's axiom that the understanding of present forces is the key to the past is now the basis of the natural sciences. Yet he was a flexible man and averse to dogmatic interpretations of his doctrine. "We are not," he added wisely, "to limit Nature with the uniformity of an equable progression." He was aware of violence and occasional spectacular occurrences in nature—he even caught a faint, far glimpse of the European ice age—but he was a child of the century of Enlightenment. "No powers are to be employed that are not natural to the globe," he wrote, disdaining the half-lit supernatural domain of the catastrophists. Such remarks cost his memory ill, later on. For Hutton, who, like Newton, was a devout man, believed in reason because reason itself lay behind nature and had directed its course. He was, in this respect, a typical eighteenth-century deist who accepted the sanctity of undeviating law and avoided the intrusion of the supernatural into the natural realm.

VI

AT THIS point we are faced once more with the question upon which I touched on an earlier page: namely, whether it is correct to interpret Hutton as viewing the earth as a living organism—what we might call the eighteenth-century physician's view —or whether his analogy is not rather that of the celestial machine of Newton. In the answer we will obtain a better glimpse into the preconceptions of the age. I hold for the machine, but it is justifiable to observe that the animal had, in some eyes, also become a machine. The two views are actually unitary.

It is true that Hutton speaks metaphorically of the body of the earth as wasting and being replenished like an animal body. Elsewhere, however, he speaks of the earth as "a machine of peculiar construction" and again he refers to this "beautiful machine."

Besides the influence of the Newtonian celestial machine which so impressed the eighteenth century, there is the powerful example of Newton's experimental method, to which Hutton clung resolutely. Hypotheses were subordinated to experiment. The Newtonian machine was one created by fiat, not one growing like an animal. Hutton seems not to have been greatly affected by evolutionary doctrines save for his acceptance of the long time scale. He believed that the world, like Newton's celestial machine, was run on perfect principles and was self-balancing rather than undergoing what we today would term unreturning, complete historicity. Hutton was also influenced by the general interest in James Watt's steam-engine experiment and is known to have spoken of volcanoes as "safety valves." His world machine sounds at times like a heat engine.

Thus it would appear in the great scientist-physician's memoirs that the world machine was variously conceived. The French experimenters of the seventeenth and eighteenth centuries had been so intrigued, since the days of Descartes, with the idea that animals were pure soulless automata that many cruel and heartless experiments had been performed, as the following contemporary account from La Fontaine attests:

> "They administered beatings to dogs with perfect indifference and made fun of those who pitied the creatures as if they had felt pain. They said that the animals were clocks; that the cries they emitted when struck, were only the noise of a little spring which had been touched, but that the whole body was without feeling. They nailed poor animals up on

boards by their four paws to vivisect them and see the circulation of the blood which was a great subject of conversation."

These historical items make plain that the organism-machine analogies were not remote from each other at that time except in so far as man, in contrast to the animal machine, had a soul. When this distinction was no longer scientifically tenable, there would emerge a genuine difference between animal and machine, because organisms form themselves and evolve, while machines do not.

Thus when the Divine Maker was retired from the earthly scene by science, leaving only secondary causes to operate nature for him, men, animals and the celestial and world machines alike were no longer to be quite what they had been in the days of supernatural intrusion, of a tampering by the Unseen. Man's world was finally to be completely natural. Yet at the close of the eighteenth century it was still a world considered to be of divine origin and created for human habitation. Only later would it be found a world without the balance of stabilized perfection. The Microcosm would not repeat the Macrocosm. The celestial clocks would no longer chime in perfect order.

Edmund Halley, as early as 1717, had calculated that the entire solar system must be moving mysteriously toward remote constellations. Man, too, was to become as natural as the wandering stars that lighted his unknown course. He was to learn that his habitation was unfixed. Not only he but his tightly governed universe was soon to be adrift and moorless on the pathways of the night.

II

HOW DEATH BECAME NATURAL

The world is the geologist's great
puzzle box.

LOUIS AGASSIZ

I T I S necessary in surveying the human quest for certainty to consider death before life. I have not done this out of perversity. Rather I have done it because, in the sequence of ideas we have been studying, it is necessary to understand certain aspects of death before we can comprehend the nature of life and its changes.

Man, even primitive man, has tended to take life for granted. Death was the unnatural thing, the result of malice or mistake, the after-message of the gods, or, in the Christian world, the result of the Fall from the Garden. In the development of a scientific approach to life on this planet, therefore, the recognition of death—species death, phylogenetic death—had to precede the rise of serious evolutionary thought. For without the knowledge of extinction in the past, it is impossible to entertain ideas of drastic organic change going on in the present or future.

Moreover, extinction is not something which can be postulated from a philosopher's armchair. It can be ascertained only by careful and precise field observation. Comparative anatomy has to be carried to a sufficient point of accuracy that the existing fauna of the world can be distinguished from the faunas of the past. The deeper our knowledge of the geological record penetrates, the stranger are the forms which can be discerned in the earth's far

epochs. Without this historical perspective any suggestion of plant or animal change is bound to be
limited and the imagination impoverished. At best
such ideas will be confined to what can be observed
in the way of change among modern products of
the breeder's art. Breeding did, however, promote
a kind of incipient evolutionism. Thus one of the
very early and anonymous commentators upon selection in England, in discussing domesticated
forms, has this to say: "Amidst these varieties,
which have sprung up under our eye, there are not
a few which deviate so much from the type of the
species, that we seem incapable of assigning a limit
to man's power of producing variation; nor when
thinking how many similar circumstances acci-
dently occur in nature, is it easy to avoid suspecting
that many reputed species may in reality have descended from a common stock." [1]

It was the lack of knowledge of the fossil past
which so greatly handicapped the first evolutionists
of the eighteenth century. It is just here that the
failure of Hutton's views to be received as credible
is disappointing. Hutton, while not a student of
fossils, was, as we have seen, a student of time. He
could read its passage in the rocks and he had been
prepared to venture a belief in the enormous an-

[1] Anonymous, "On Systems and Methods In Natural History,"
The Quarterly Review (London) , Vol. 41 (1829) , p. 307. On
the basis of interior evidence I am inclined to suspect that this
paper is a youthful and unacknowledged review by Sir Charles
Lyell.

tiquity of the earth. There can be no doubt that the conservative English reaction in science after the French Revolution delayed the recognition of Huttonian geology. As an indirect consequence, it may well be that the acceptance of the evolutionary philosophy itself was also delayed by a generation. Time and accompanying geological change are two of the necessary properties without which evolution would be unable to operate. And those two properties bring death as a third factor in their wake.

I I

THE seventeenth century was, in general, still in the grip of the short Christian time scale, though here and there, toward the close of the century, some hesitant doubts began to be expressed by men like John Ray. It was the general opinion, says Ray, "among Divines and Philosophers that since the first creation there have been no species of animals or vegetables lost, no new ones produced." It was part of the reigning theology of the time that extinction was an impossibility. In fact, this view continued to be reiterated in the eighteenth century and, by ultraconservative thinkers, into the first decade of the nineteenth century.

Thomas Jefferson, writing in 1782, commented that "such is the economy of nature that no instance can be produced of her having permitted any one race of her animals to become extinct; of her having formed any link in her great work so weak as to be broken." The well-read Jefferson is, of course, merely repeating what was the commonly accepted view of his time. Extinction loomed as something vaguely threatening and heretical. In fact, for that very reason many refused to accept fossils as representing once-living creatures.

Their reluctance to accept what now seems to us so easily discernible and commonplace an observation as extinction is based essentially upon one fact: the benignity of Providence. "To suppose any species of Creatures to cease cannot consist with the Divine Providence," writes one seventeenth-century naturalist, and his comment is frequently reiterated by others. This point of view is based upon a theory of organic relationships which, though traceable into earlier centuries, reached a peculiar height of development during the seventeenth and eighteenth centuries. The belief was not confined to philosophy and theology. It permeated the whole field of letters and became widely known as the Scala Naturae, or Ladder of Being.

This conception superficially resembles a line of evolutionary ascent and undoubtedly has played an indirect part in the promotion of evolutionary notions of a scale of rising complexity in the development of life. It was not, however, a scheme of evolution. It is based on a gradation which emerged instantaneously at the moment of creation, and which rises by imperceptible transitions from the inorganic through the organic world to man, and even beyond him to divine spiritual natures.

The idea promoted much anatomical work of a devout character as naturalists attempted to work out the missing or obscure links in what was regarded as an indissoluble chain which held together the various parts of creation. It was for this reason that men viewed with genuine horror the idea that links could be lost out of the chain of life. The strong belief in an all-wise Providence which did nothing without intention caused men to refuse an interpretation of the universe which involved apparently aimless disappearances. If such disappearances were possible, what might not happen to man himself?

Just as there existed the balanced, self-correcting machine of the heavens, and the balanced, self-renewing machine of the earth, so life was similarly linked in the great chain of unalterable law. All was directly under the foreknowing care of the Divine Being. Nevertheless, by the early eighteenth century it began to be whispered among English naturalists "that many Sorts of Shells are wholly

lost, or at least out of our Seas." The simpler organisms from marine strata were being identified
before the taxonomy of fossil vertebrates had been
seriously attempted.

III

It MAY now be asked, as fossils slowly became accepted as the remains of creatures once living, how
men were for so long able to evade the still troubling question of extinction—the existence of
death before the Garden. A world view does not
dissolve overnight. Rather, like one of Hutton's
mountain ranges, it erodes through long centuries.
In the case of fossils we must remember how small
the European domain of science was in comparison
with the vast continental areas which had only recently been opened for examination by the voyagers. Australia, Africa, the Americas, were barely
known; their interiors remained unexplored. They
provided a providential escape for the devout naturalist who still wished to avoid the dangerous logic
implicit in unknown bones and shells.

Since it was beginning to be realized that the
separate continents possessed faunas and floras to
some degree distinct, the devout could argue that
creatures of which there were now no living repre-

sentatives in Europe might well have been driven out by man, or by changes of climate; in short, that instead of being extinct, they had merely retired to the fastnesses of unknown seas or continents. Jefferson quotes an observer who had heard the mammoth roaring in the Virginia woods. In Europe the bones of Ice Age elephants were ascribed to the living African species imported, so it was claimed, for Roman games. Or the bones were those of Hannibal's war elephants lost on the Alpine passes. The desired point was to make the animals *historic* and thus ascribable still to living species. In America, where great bones lay in profusion, it was rumored that living specimens could be found across the unknown Great Lakes or farther on in the heart of the continent. By the mid-eighteenth century the aroused pursuit of natural curiosities reveals the fascination of a public increasingly alive to foreign rarities and the new mysteries of the living world.

Peter Collinson, the Quaker merchant and naturalist in London, writes testily to John Bartram in the colonies: "If thee know anything of thy own knowledge please to communicate it. The hearsay of others can't be depended on."

Bartram in turn complains peevishly, "The French Indians have been very troublesome, which hath made travelling very dangerous beyond our [territory] where I used to find many curiosities.

. . . While we . . . daily expect invasions we have little heart or relish for speculations in Natural History."

Still, in spite of Indians and the depredations of pirates, the letters and the specimens made their slow way to Europe. "The frogs came safe, and lively. I transcribed thy account of them, and had it delivered to the King."

"I received the turtle in good health; and shall be much obliged to him if he will procure me a male and female Bull-frog. Mine are strayed away."

"Your ingenious idea respecting the former existence of certain kinds of animals, now extinct, I confess carries great weight with it and yet, my dear Sir, I cannot implicitly give my assent to it on the whole. With regard to the Unicorn I am rather divided in my judgment, even in respect to their present existence, in the interior region of Africa, of which, we are extremely ignorant."

"That all petrifactions should be attributed to the general deluge, is what I shall never agree," growls another correspondent taking a slight step toward the future.

Reading the old letters, we hear the voices mingle in a mounting symphony. "Every uncommon thing thou finds in any branch of Nature will be acceptable." "The terrapins had bad luck. Some died, others the sailors stole." "I had some doubts, so carefully examined the Ohio Elephant's long teeth with a great number . . . from Asia and from Africa and found they agree with what is called the *Mammot's* Teeth from Siberia. It is all a wonder how they came to America. . . ."

Then Collinson turns to the as yet unknown teeth of the mastodon and the extinct Irish elk. "So here are two animals, the creature to which the great forked or pronged teeth belong [mastodon]. Whether they exist, God Almighty knows,—for no man knows: whether ante-diluvians, or if in being since the flood. *But it is contrary to the common course of Providence to suffer any of his creatures to be annihilated.*"

In Germany one ingenious escapist propounded a theory that the rocks which compose the geological strata of the earth had fallen at various times from the heavens as meteoric or planetary fragments. Thus, he contended, the unknown species of animals whose fossils had now become so troubling to the devout had no necessary connections with this earth at all. They were the remains, instead, of extraterrestrial creatures. It was a disengaging action almost two hundred years ahead of the space age, which would have been fascinated with such a notion.

We are now in position to see that the eighteenth

century—powerfully rationalistic and scientifically curious though it was—had difficulty until toward the final decade of the century, in assimilating the idea that species could be utterly extinguished. Deism was the reigning philosophy of the times, and deism repudiated the idea that God immediately interposed his will in nature. Rather, he delegated powers to secondary causes which were self-operating.

We have seen this view expressed in the heavenly and earthly machines together. The deists, however, along with other rationalists, reposed great faith in human reason as reflecting the Divine reason. God, it was thought, could be known through nature and apart from Biblical revelation. The problem which caused such hesitation among the eighteenth-century students of animal life, therefore, was how to explain the apparent irrationality and waste involved in the discovery of extinction. Why would a supremely rational God reject and repudiate his creations? Furthermore, if such repudiation occurred, was there not danger that man himself might be swept from the stage of life?

Secondary forces as controlling the world, man had come to accept. He had passed beyond the conception of a God supernaturally intervening in mundane affairs. Nevertheless, man had continued to believe that the great machine was rationally organized with human welfare in mind—that it was self-balanced and its aberrations self-correcting. The hint of extinction in the geological past was like a cold wind out of a dark cellar. It chilled men's souls. It brought with it doubts of the rational world men had envisaged on the basis of their own minds. It brought suspicions as to the nature of the cozy best-of-all-possible-worlds which had been created specifically for men.

A vast and shadowy history loomed in the rocks. It threatened to be a history in which man's entire destiny would lose the significance he had always attached to it. For a few decades the lost links of life might be sought as living beyond the sources of primeval Amazons and Orinocos. In the end man's vision of his world would undergo drastic revision. Out of it would emerge once more, though briefly, a renewed confidence in his position in the universe.

IV

WE HAVE already had occasion to observe the preference in man to interpret startling features of landscape in terms of catastrophic violence. In the years immediately following Hutton's death, the slow alteration of earth's features postulated by him and the Frenchman Lamarck gave way before the widespread popularity of the geological doctrine of catastrophism. Linked with it, there appeared a new and highly popular theory in which extinction was finally recognized by science. It was, however, explained in a manner that was pleasing to the public fancy and at least offered a compromise between science and older theological beliefs.

"At certain periods in the development of human knowledge," C. D. Broad once remarked, "it may be profitable and even essential for generations of scientists to act on a theory which is philosophically quite ridiculous." This was true in a comparatively short-lived way of catastrophism. It persuaded man to accept both death and progressive change in the universe. It did so by extending such mythological events as the world-shattering Biblical Deluge, and by the creation of a form of geological prophecy which left man still the dominant figure in his universe.

"Half a century ago," wrote the great American botanist Asa Gray, who died in 1880, "the commonly received doctrine was that the earth had been completely depopulated and repopulated over and over; and that the species which now, along with man, occupy the present surface of the earth, belong to an ultimate and independent creation, having an ideal but no genealogical connection with those that preceded. This view . . . has very recently disappeared from science."

This was the philosophy which Sir Charles Lyell

was destined to overthrow; this was the view propounded to young Charles Darwin by his geology
professors before he went on the voyage of the
Beagle. This was the concept against which Darwin
dueled with Agassiz past midcentury. In what lay
the vitality of this weirdly irrational theory, and
how did it arise? We have hinted at its appeal to
human vanity in a time of growing religious confusion. To examine its roots we must turn again to
the closing years of the eighteenth century and the
first two decades of the nineteenth. In those years
we find a growing amalgamation of new ideas as
follows:

1. Cuvier and his associates, working upon the
successive vertebrate faunas of the Paris basin, remarked "that none of the large species of quad-
rupeds whose remains are now found imbedded in
regular rocky strata are at all similar to any of the
known living species." Furthermore, they denied
that the species concerned "are still concealed in
the desert and uninhabited parts of the world."

2. Cuvier could observe no graduated changes in
his vertebrate fossils. He assumed, therefore, that
no intermediate forms existed between one geological level and another. That there was a progression in the complexity of the animal types from age
to age as we approach the present, Cuvier had,
however, begun to perceive.

3. Cuvier assumed that there were genuine

breaks between one age and another brought about by catastrophic floods and alterations in the relation of sea to land. He did not, however, propose such a total break between one age and another as the later catastrophists.

4. There arose among German and French nature philosophers a renewed sense of the unity of plan or biological structure common to large groups of plants and animals—a morphological advance also marked by the contributions of Cuvier in France and his disciple, Richard Owen, in England. The transcendental aspects of this morphology lay in the conception that these major structural plans existed abstractly in the mind of God, who altered them significantly from age to age. In the words of Adam Sedgwick, Darwin's old teacher, "At succeeding epochs, new tribes of beings were called into existence, not merely as the progeny of those that had appeared before them, but as new and living proofs of creative interference; and though formed on the same plan, and bearing the same marks of wise contrivance, oftentimes as unlike those creatures which preceded them, as if they had been matured in a different portion of the universe and cast upon the earth by the collision of another planet."

It was generally conceived that the lower and earlier forms of life pointed on directly to man, who had been ordained to appear since the time of the first creation. "It can be shown," asserted Louis Agassiz, "that in the great plan of creation . . . the very commencement exhibits a certain tendency toward the end . . . The constantly increasing similarity to man of the creatures successively called into existence, makes the final purpose obvious . . ." The geological record was being searched for signs and portents pointing to human emergence at a later epoch. There is more than a hint of medieval "signs in the heavens" to be found in these paleontological auguries: a reptile leaving handlike imprints on some ancient sea beach is a portent of man's coming; the stride of a bipedal dinosaur discloses the eventual appearance of bipedal man.

Catastrophism, if we are to examine it in its most mature form—that of England in the second decade of the nineteenth century—has, as we have seen, several surprising features. The deathless Eden of the Biblical first creation has been replaced by a succession of natural but successive worlds divided from each other by floods or other violent cataclysms which absolutely exterminate the life of a particular age. Divinity then replaces the lost fauna with new forms in succeeding eras. Disconformities in geological strata, breaks in the paleontological record, are taken as signs of world-wide disaster terminating periods of calm. In contrast to eighteenth-century concern over the death of species, and anxiety to establish seemingly extinct animals as still in existence, the natural theologians now revel in violence as excessive as that of the Old Testament. Whole orders of life are swept out of existence in the great march toward man. The stage which awaits the coming of the last great drama has to be prepared. Floods destroy the earlier actors. Enormous death demands equally enormous creation, discreetly veiled in the volcanic mists that hover over this half-supernatural landscape.

From the idea that one lost link in the chain of life might cause the whole creation to vanish piecemeal, man had passed, in scarcely more than a generation, to the notion that the entire world was periodically swept clean of living things. The discoveries of vertebrate paleontology seemed to illuminate, in solitary lightning flashes, a universe that progressed in leaps amidst colossal destruction.

Still, there was a pattern amidst the chaos. For while strange animals arose and perished, it was observed that the great patterns of life, the divine blueprints, one might say, persisted from one age to another. It was only the individual species or genera that vanished and each time were re-created in an altered pattern. There was no natural connection, no phylogeny of descent in a modern sense. There was only the Platonic ideal of pure substanceless form existing in the mind of God. The reality of material descent escaped the mind of the observer.

The public concentrated less upon living animals adjusting to circumstance than upon this strange spiritual drama which the natural theologians had read into the rocks. Two steps in the direction of naturalism had been gained, however. First, the world of the past was now known to have cherished plants and animals no longer to be observed among the living. Second, thanks largely to the pioneer efforts of William Smith of England, who had once caustically remarked to critics that "the search for a Fossil may be considered at least as rational as the pursuit of a hare," it was known that fossils could be used to identify distinct strata.

Smith had recognized the changes in invertebrate fossils from one geological level to another. Cuvier's observations upon vertebrate remains had similarly, if more dramatically, established a type of grand progressive movement among the vertebrates. Life did not return upon its track. The record in the book of stone showed no reversals. Life, in other words, was a historic progression in which the past died totally. But the goal was finalistic—it was man. Even coal forests had been laid down for his use. At times it seemed that the earlier creation existed only as some kind of phylogenetic portent of man.

V

CATASTROPHISM, in essence, may be said to have died of common sense. As a modern historian, Charles Gillispie, has commented, "To imagine the Divine Craftsman as forever fiddling with His materials, forever so dissatisfied with one creation of rocks or animals that He wiped it out in order to try something else, was to invest Him with mankind's attributes instead of the other way about."

Slowly the accumulation of geological information began to lead back toward the pathway pursued earlier by James Hutton and his follower, John Playfair. Sir Charles Lyell, who was born the year of Hutton's death, reapproached the whole subject of uniformitarian geology in his famous *Principles of Geology,* whose first volume was published in 1830. In the intervening half century, much additional information had become available. Lyell was a careful organizer of facts, a man of judicious temperament and an independent thinker. He, like Hutton, had an eye for the common observable workings of sunshine and water drops. In fact, he was one of the first to read in fossil impressions that the raindrops of the past were similar to those of the present, that the eyes of fossil trilobites showed light falling upon the earth many millions of years ago as it falls today. He saw no evidence of world-wide catastrophes. He observed, instead, local disconformities of strata, the rise and fall of coast lines, the slow upthrust of mountain systems. He saw time as illimitable, in the fashion of Hutton.

Moreover, Lyell attempted statistical estimates of the change in molluscan species as one passed from one distinct bed to another. He observed that certain organisms persisted—though in changed proportions. He could not discover the drastic and sudden eliminations of fauna upon which the catastrophists had built their case. As work progressed, more and more of what were termed "passage beds" appeared—strata linking one supposedly separate era with another. Local sequences of this sort began to make clear the essential unity of earth's geological history. By degrees Lyell's more vociferous opponents grew silent. In that silence one thing was clearly apparent. What we might call point-extinction, i.e., extinction of the individual species, had replaced the concept of mass death. Death, in other words, was becoming natural—a product of the struggle for existence.

Lyell observed that the long course of geological change was bound to affect the life upon the planet's surface. He saw that every living creature competed for living space and that every change of season, every shift of shore line, gave advantages to some forms of life and restricted space available to others. Over great periods of time it was inevitable that some species would slowly suffer a reduction in numbers and, by degrees, perish, to be replaced by others.

This incredibly tight and complicated web of life would, Lyell thought, eliminate immediately any newly emerging creatures which might be evolving through natural means. Yet if, as the geological record indicated, species perish, somewhere there must be creation, somewhere there must be a coming in to replace the deficit involved in extinction.

Lyell hesitated. Could there be individual point-emergences as well as vanishings? No one professed to have seen the creation of a complex animal. Was it because so rare a thing was simply not normally observable?

Lyell had inherited from Hutton a distaste for the unseen. He preferred to work with visible, understandable forces susceptible of observation. He did not like catastrophism with its spectacular and unobservable creations any more than he liked its flamboyant geological mechanisms. As early as 1829 he had written privately, "We shall very soon solve the grand problem, whether the various living organic species came into being gradually and singly in insulated spots, or centers of creation, or in various places at once, and all at the same time. The latter cannot, I am already persuaded, be maintained."

Try as he might, Lyell could find no satisfactory explanation for the advances in biological organization which the catastrophists acclaimed. To admit them was to accept miracle—the unknown. To accord them acceptance was equivalent to geological capitulation also—it was, in effect, to say: "I cannot explain your mysterious and advancing creation of life, therefore if life is miraculous, your interpretation of geological forces may just as well be accepted in the same fashion."

At this Lyell balked—not always consistently. He pointed out that the geological record was incomplete. He contended that the discovery of vertebrates in older deposits than they were originally assumed to characterize was a sign of nonprogression, that the serial advances recorded by the progressionist-catastrophist school were largely in error.

Faunas might shift with time and geography, Lyell warned, but this might not involve necessary progression through the vertical realm of geology. Only man Lyell admitted to be young. The extinct forms amidst the great phyla could be accepted without the argument that there was a common, necessary upward trend in creation culminating in man.

Up to the time of the publication of the *Origin of Species,* Lyell was suffering from the lack of a satisfactory explanation of organic change. He had overthrown the extinction-in-mass conception of the catastrophists; he had reintroduced into geology the lengthier time span of Hutton, and Hutton's devotion to purely natural forces. It was, however, difficult to see how those forces applied to the single great mystery—life.

Lyell stood, actually, at the verge of Darwin's discovery. He was, however, within the shadow of another century—the eighteenth. He shared its passion for intellectual order, for the obedient and unmysterious world machine. Hutton had taken life for granted because almost nothing was known in his day of the antediluvian world of fossils. Lyell, by contrast, was confronted by a perverse, unexplainable force that crawled and changed through the strata—life. He made *death* natural, but it could be said that life defeated his efforts to understand it. With all their errors, the catastrophists had been right about one thing. From its early beginnings in the seas, life *had* been journeying and growing in complexity. It is historic.

All the way back into Cambrian time we know that sunlight fell, as it falls now, upon this planet. As Lyell taught, we can tell this by the eyes of fossil sea creatures such as the trilobites. We know that rain fell, as it falls now, upon wet beaches that had never known the step of man. We can read the scampering imprints of the raindrops upon the wet mud that has long since turned to stone. We can view the ripple marks in the sands of vanished coves. In all that time the ways of the inanimate world have not altered; storms and wind, sun and frost, have worked slowly upon the landscape. Mountains have risen and worn down, coast lines have altered. All that world has been the product of blind force and counterforce, the grinding of ice over stone, the pounding of pebbles in the mountain torrents—a workshop of a thousand hammers and shooting sparks in which no conscious hand was ever visible, today or yesterday.

Yet into this world of the machine—this mechanical disturbance surrounded by desert silences —a ghost has come, a ghost whose step must have been as light and imperceptible as the first scurry of a mouse in Cheops' tomb. Musing over the Archean strata, one can hear and see it in the subcellars of the mind itself, a little green in a fulminating spring, some strange objects floundering and helpless in the ooze on the tide line, something beating, beating, like a heart until a mounting thunder goes up through the towering strata, until no drum that ever was can produce its rhythm, until no mind can contain it, until it rises, wet and seaweed-crowned, an apparition from marsh and tide pool, gross with matter, gurgling and inarticulate, ape and man-ape, grisly and fang-scarred, until the thunder is in oneself and is passing—to the ages beyond—to a world unknown, yet forever being born.

"It is carbon," says one, as the music fades within his ear. "It is done with the amino acids," contributes another. "It rots and ebbs into the ground," growls a realist. "It began in the mud," criticizes a dreamer. "It endures pain," cries a sufferer. "It is evil," sighs a man of many disillusionments.

Since the first human eye saw a leaf in Devonian sandstone and a puzzled finger reached to touch it, sadness has lain over the heart of man. By this tenuous thread of living protoplasm, stretching backward into time, we are linked forever to lost beaches whose sands have long since hardened into stone. The stars that caught our blind amphibian stare have shifted far or vanished in their courses, but still that naked, glistening thread winds onward. No one knows the secret of its beginning or its end. Its forms are phantoms. The thread alone is real; the thread is life.

"Nevertheless, there is a goal," we seek to console ourselves. "The thread is there, the thread runs to a goal." But the thread has run a tangled maze. There are strange turns in its history, loops and knots and constrictions. Today the dead beasts decorate the halls of our museums, and that nature of which men spoke so trustingly is known to have created a multitude of forms before the present, played with them, building armor and strange reptilian pleasures, only to let them pass like discarded toys on a playroom floor. Nevertheless, the thread of life ran onward, so that if you look closely you can see the singing reptile in the bird, or some ancient amphibian fondness for the ooze where the child wades in the mud.

One thing alone life does not appear to do; it never brings back the past. Unlike lifeless matter, it is historical. It seems to have had a single point of origin and to be traveling in a totally unique fashion in the time dimension. That life was ever a fixed chain without movement was a human illusion; that it leaped as some mystical abstraction from one giant scene of death to another was also an illusion; that geological prophecy proclaimed the coming of man as Elizabethan astrologers read in the heavens the signs of coming events for kings was an even greater fantasy. Instead, species died irregularly like individual men over the long and scattered waste of eons. And as they died they must, as Lyell foresaw, be replaced in as scattered a fashion as their deaths. But what was the secret? Did a voice speak once in a hundred years in some hidden wood so that a nocturnal flower bloomed, or something new and furry ran away into the dark?

Creation and its mystery could no longer be safely relegated to the past behind us. It might now reveal itself to man at any moment in a farmer's pasture, or a willow thicket. By the comprehension of death man was beginning to glimpse another secret. The common day had turned marvelous. Creation—whether seen or unseen—must be even now about us everywhere in the prosaic world of the present.

III

HOW LIFE BECAME NATURAL

---

*If we can conceive no end of space,
why should we conceive an end of
new creations, whatever our poor
little bounds of historical time
might even appear to argue
to the contrary.*

LEIGH HUNT, 1836

GREAT literary geniuses often possess an ear or a sensitivity for things in the process of becoming, for ideas which are just about to be born. It is interesting in this connection to compare the remarks of Charles Darwin with certain observations on science made at a much earlier date by the poet Samuel Taylor Coleridge. Darwin, in his autobiography, protests that he saw no evidence that the subject of evolution was "in the air" of his time. "I occasionally sounded out not a few naturalists," he remarks, "and never happened to come across a single one who seemed to doubt about the permanence of species."

By way of contrast, we may note that Coleridge, in a philosophical lecture delivered as early as 1819, makes reference to a belief which "has become quite common even among Christian people, that the human race arose from a state of savagery and then gradually from a monkey came up through various states to be man." Coleridge was not an evolutionist. He is, however, sensitive to a new doctrine, whose presence "in the air" Darwin had failed to discover. He observes in a very shrewd fashion what we have sought to emphasize in previous chapters; namely, the way in which the intellectual climate of a given period may unconsciously retard or limit the theoretical ventures of an exploring scientist. "Whoever is acquainted with the history of philosophy during the last two or three centuries," contended the great poet, "cannot but admit, that there appears to have existed a sort of secret and tacit compact among the learned, not to pass beyond a certain limit in speculative science. The privilege of free thought so highly extolled, has at no time been held valid in actual practice, except within this limit."

Coleridge is here recognizing a fact which escaped the simpler and less philosophically oriented mind of Darwin. The latter had failed to recognize that the silence of his professional colleagues was in some cases just such a restriction as that of which Coleridge speaks. Religious and social pressures had all contributed to making the subject of evolution somewhat taboo and not really in good taste to discuss in public. Time and again one finds naturalists circling all about the subject and then withdrawing timidly from any attempt to derive the final, and what now seems to us the logical, conclusion. Some, in fact, openly contradict themselves in order to stay within the accepted circle of traditional thought.

The entomologist T. Vernon Wollaston is a case in point. In his book *On the Variation of Species,* published in 1856—a volume from which Darwin drew material favorable to the evolutionary point of view—Wollaston dwells sporadically upon organic change. We are led to believe, he says, "that, could the entire living panorama, in all its magnificence and breadth, be spread before our eyes, with its long lost links [of the past and present epochs] replaced, it would be found, from first to last, to be complete and continuous throughout,—a very marvel of perfection, the work of the Master's Hand. . . . From first to last," he contends, "the same truth is re-echoed to our mind, that here all is change." Yet after making these and other suggestive remarks of a similar character, Wollaston disavows in his final chapter the idea of the transmutation of species, though at the same time he can only say that the limitation of a species must be indicated *somewhere*. This timidity, which fits so well Coleridge's observations of the scientific mind, savors more of religious discomfort than genuine scientific conviction. It characterizes a considerable amount of the biological literature of the earlier half of the nineteenth century.

In the end it was an outsider, Robert Chambers, without professional ties, an amateur, as again Coleridge was sharp enough to generalize upon, who would break through the orthodox barrier of conservatism. Finally, it would be another wealthy amateur, Charles Darwin, whose massive treatise would swing world thought into a new channel. The story is a fascinating one. Coleridge's critical observations upon science made forty years before the publication of the *Origin* are an almost exact preview of what happened in the long history of the evolutionary concept.

The period prior to Darwin's enunciation of the full evolutionary doctrine is difficult to define with precision. Just as in the orchestra pit before a great musical performance there is the individual tuning of strings, plucking of stray notes, and discordant thumpings before the conductor, with his baton, brings unity into a composition, so in the period prior to 1859, we can get all manner of tentative retreats and approaches, partial developments, hesitant insights and bold sounds from the wings. When, at last, Darwin picks up the conductor's wand and turns this assemblage of stray notes into a full-throated performance, the world audience is swept off its feet so completely that, hypnotized by Darwin's single figure, it forgets the individual musicians who made his feat possible.

In actuality, life was not made natural in a day, nor in a single generation. Neither, we have observed, were geology and death. Before entering upon the difficult problem of life, therefore, let us now attempt to recall the major shifts which had taken place in scientific thinking upon this subject. It will be remembered that the seventeenth, and to a very considerable extent the eighteenth, century believed in a linked, unbreakable chain of organisms ascending in complexity to man. In spite of the fact that this conception bears a superficial resemblance to the idea of evolution, it is, in reality, a fixed, static, and immovable chain. Nothing changes position, nothing alters, nothing becomes extinct. By degrees, as the rock strata were found to contain organisms unlike those of the present era, two new conceptions arose. One of these came to be called progressionism and is associated with the catastrophic geology which we have previously examined.

This theory assumes that life was first manifested on the planet by simple organisms, but that from the very beginning of time, the stage was being set for man. Each ascending fauna, as we have seen, was successively swept away at the close of a catastrophic geological episode. For reasons unknown to man, but nevertheless mysteriously prophesied in the rocks, this lengthy prologue had preceded the human emergence. Each fauna, however, in spite of its anatomical relationship to the previous order, represented a separate act of creation. A kind of ideal Platonic morphology had been substituted for the notion of a direct physical descent which links the creatures of one age and those of the next.

The second idea which, although originally regarded askance, particularly in England, was being whispered about toward the close of the eighteenth century was that of evolution itself—actual physical descent with bodily alterations from one age to another. The cosmic evolutionism which had begun to enter the speculations of the astronomers had now attracted some inquiring minds among the biologists. These early approaches to the problem were, however, vitiated by a certain casualness and lack of evidence. The age of the world was still being underestimated, and the lack of evidence for any really marked extinction (the notion of evolution actually preceded full-blown progressionism) led to an underestimation of the possibilities contained in the idea. Lamarck, who wrote the most extended treatise upon the subject, never was very sure about the matter of extinction. He succeeded in largely evading the subject by assuming that the missing animals had evolved, without perishing, into other forms. In skirting the problem of extinction, he was a typical child of the eighteenth century. Nevertheless, it must be recognized that Lamarck, besides grasping the reality of the evolutionary process, observed that creatures fitted themselves to the environment they occupied, rather than being made for that specific environment. In this respect alone Lamarck was years ahead of his time, because until geological change has been wedded to organic change, one cannot have a full-fledged evolutionary theory.

It has occasionally been said that Lamarck remained unappreciated because he entertained some ideas which sound ridiculous to the modern ear. Historical hindsight in such matters is rarely unprejudiced. One might as well argue that Newton should have passed unheeded because he wrote lengthily upon theology, or because he manifested paranoid mental tendencies in his declining years. If we were to ignore certain of our scientific forerunners upon such a basis, we would have to dismiss the discoveries of many geniuses of the scientific twilight who entertained advanced notions along with a sincere belief in witches. In time, scientific historians looking back will undoubtedly see our beliefs as shot through and through with the equivalent fantasies of our own age. It is not just to dismiss Lamarck on such a basis, for if we were to catalogue each change in thought that led on to Darwin, we would have to recognize that this much maligned French thinker glimpsed ecological change and adjustment before Darwin. In the process he recognized what Darwin was later to call the "law of divergence" and what the modern world calls "adaptive radiation."

At this moment of recognition, just as Halley had unwittingly set the solar system adrift without a pilot, so Lamarck, though he did not realize it, had destroyed the preordained character of the human emergence. Beside such a momentous observation, the question of whether his evolutionary mechanism was right or wrong lapses into comparative unimportance. The fact that generations of historians have seen this man purely as an advocate of a now rejected explanation of how evolution comes about is the result of two misfortunes: the fact that he was a Frenchman who survived the Revolution, and the additional misfortune that his successor, Charles Darwin, had little interest in, or concern with, the history of the subject. Darwin also cherished the common conservative English attitude toward the thinkers of revolutionary and post-revolutionary France. Something of a conspiracy of silence surrounded Lamarck's name and English naturalists disavowed his theories with almost ritualistic fervor.

Our first step in the effort to understand how life became natural, therefore, is to avoid the commonly held impression that Darwin, by a solitary innovation—natural selection—transformed the western world view. Without detracting in the least from his importance, we may observe that few, if any, scientific discoveries are made in such a fashion. Newton once made the perceptive observation that if he saw far, it was because he stood on the shoulders of giants. Similarly, Charles Darwin was the inheritor of the efforts of his forerunners, but because of a new twist which he gave to those same efforts, the stages in the process leading to Darwin's achievement have passed largely unobserved. The drama of the voyage of the *Beagle,* the isolation of the years at Downe, the great shift in public opinion which began with the final acceptance of geological time through the studies of Lyell—all, in a sense, have obscured rather than illuminated the Darwinian story.

Darwin has been left in solitary grandeur as a kind of psychological father figure to biologists. Let me repeat at this point, since I have already experienced the amount of emotional heat which can still be generated about this man, that I am interested only in the presentation of a succession of some easily verifiable ideas and a view of how they changed in order to bring us to the world we inhabit today. Darwin, some biologists have proclaimed, had nothing to do with his forerunners. Others, similarly aroused, have persisted in the reverse argument that Darwin had certainly made use of the ideas of his forerunners, but that this did not matter in the least because Darwin was the man who brought the public to a recognition of evolution.

Both of these remarks strike one as emotionally oriented. They serve to conceal a certain type of unsophisticated hero worship which still exists among occasional scientists unfamiliar with the history of ideas. What we are interested in at this point is solely how an idea, natural selection, beginning as a conservative eighteenth-century observation, was altered by slow degrees into something which set the world of life adrift in an

unfixed wilderness as surely as Halley's sensitivity
had set the star streams pouring through unimaginable darkness and distance.

II

FOUR propositions, it can now be observed, had to
be clarified before the theory of organic evolution
would prove acceptable to science. *First,* as we have
already noted, the great antiquity of the planet had
to be grasped. Otherwise life would occupy too
narrow a segment of time for change of a slow nature to be possible. In fact, prior to the time of
Linnaeus, abrupt spectacular mutations were occasionally discussed in agricultural works—although without reference to evolutionary ideas.
The notion that species were completely immuta-
ble seems to have come in with a hardening of the
religious temper, particularly in the century between about 1750 and 1859.

Second, it was necessary to establish the fact
that there had been a true geological succession of
forms on the planet. Though again, as we have already observed, this did not lead immediately to
the acceptance of the evolutionary hypothesis, it
did call attention to a totally forgotten series of
worlds stretching into the remote past. Inevitably
any rational philosophy would have to account for

the inhabitants of former times and, if possible, relate them to the plants and animals of the present. Knowledge of the vertebrate succession began to emerge only after 1800. As a consequence of this gap in our knowledge, paleontology for a time contributed to the growth of the progressionist scheme of mass extinctions and re-creations of life.

Third, the amount of individual variation in the living world and its possible significance in the creation of change had to be understood. Variation began to be noted and speculated about as far back as the seventeenth century, but its role in evolution would not be understood until much later. Nevertheless, the activities of stock breeders would engender some notion of at least limited organic change.

Fourth, the notion of the perpetually balanced world machine, which had been extended to life itself, had to give way to a conception of the organic world as not being in equilibrium at all—or at least being only relatively so—a world whose creations made and transformed themselves throughout eternity. Life had to be seen, in the apt phrase of a later evolutionist, Alfred Russel Wallace, as subject to "indefinite departure"— alteration, in other words, subject to no return.

We have already had occasion to examine what occurred to illuminate the understanding of time and animal succession. It is with the last two of these four propositions, variation and the balanced world machine, that we will now concern ourselves. Before life could be viewed as in any way natural —and it is not my intention to push this word too far—a rational explanation of change through the ages had to be proposed. In a sense, it was Hutton and Lyell's problem of earth change reapplied to the problems of life. It embodied a similar search for natural causes at work in the present day— causes still capable of study and observation. Variation, selection, the struggle for existence, were all known before Darwin. They were seen, however, within the context of a different world view. Their true significance remained obscured or muted in precisely the manner that Coleridge had antici- pated in his estimate of the scientific mind.

It was not really new facts that were needed so much as a new way of looking at the world from an old set of data. A few men had tried to accomplish the task even before the close of the eighteenth century. The lack of knowledge of the fossil past, however, made their attempts impoverished and undramatic compared with the catastrophism which relegated their efforts to obscurity. They had failed to supply a satisfactory explanation for evolutionary change. In addition, England was swept by an anti-French wave of conservatism which was the intellectual product first of the Revolution, and then of the following Napoleonic wars. Her science, to a degree, became isolated from that of the Continent.

III

THE historian who examines with care the documents of the eighteenth century before the recognition of extinction, and while the scale of nature is still the overriding biological as well as theological concept, will come immediately upon a principle of balance which was believed to prevail throughout the living world. This is what I meant when I said that the eighteenth-century love of order had been extended to the living world. Something of the complexity of the living environment had been observed, something of what the biologist of today would call food chains. To the eighteenth-century mind, however, this world was in a permanent, rather than dynamic, balance. It was a manifestation, in other words, of divine rule and government. Let me explain this subtlety, if I can. To do so, let us examine the work of just one man, John Brückner, a Frenchman. Brückner is the actual forerunner of Thomas Malthus, the man who so profoundly influenced the political and biological thinking of the nineteenth century.

Brückner's book entitled *A Philosophical Survey of the Animal Creation* appeared in English in

1768, over thirty years before Malthus' *Essay on the Human Population.* In it he observes, "Providence seems to have advanced to the utmost verge of possibility in the gift of life conferred upon animated beings." Famine, pestilence and war, the checks upon human population to which Malthus devoted so much attention, occur in Brückner's pages. The struggle for existence is very clear to him: "For it is with the animal as with the vegetable system: the different species can only subsist in proportion to the extent of land they occupy; and wherever the number of individuals exceeds this proportion, they must decline and perish."

Brückner, however, is still obsessed with the short Christian time scale. "It is five thousand years at least," he observes, "that one part of the living substance had waged continual war with the other, yet we do not find that this Law of Nature [i.e., natural selection] has to this day occasioned the extinction of any one species." Here Brückner is quite obviously laboring under the providential belief that extinction is an impossibility. In the next sentence Brückner makes very clear the eighteenth century's recognition of the role of natural selection. "Nay," he says, continuing his discussion of the war of species against species, "it is this which has preserved them in that state of perpetual youth and vigor in which we behold them. . . .

"The effects of the carnivorous race," he goes on

to add, "are exactly the same as that of the pruning-hook, with respect to shrubs which are too luxuriant in their growth, or of the hoe to plants that grow too close together. By the diminution of their number, the others arrive at greater perfection." Brückner calls this process "reciprocal attrition." It will take other names before it finally reaches Darwin. Innumerable times it will be referred to by scientific writers as "pruning," "policing," "natural government."

The phrase "natural government" best expresses the eighteenth-century world view of the interlinked web of life. Like Hutton's world machine, its momentary aberrations were self-correcting. Hutton himself, in an unpublished work only recently made available, reveals a clear knowledge of natural selection in this varietally selective sense —proving once more, if proof were needed, that the principle antedates Darwin. Struggle, it was thought, adjusted the quantity of life and eliminated the unfit. Beyond this, selection did not create. Life was held in a static balance. It was not going anywhere.

One other curious phrase in Brückner's work demands attention, for it, too, is indicative of the time. "If the absolute number of inhabitants is not so great as it might have been, it is nevertheless always approaching toward its plenitude." The word "plenitude" had a very special meaning to the men of this age. It was assumed that God was creative up to the limit of his capacity, that every Platonic form or idea must be expressed in a rational universe. Thus the philosophic notion of plenitude contributed abstractly to what Professor Lovejoy has called a "Malthusian picture of a Nature overcrowded with aspirants for life."

The discovery of extinction did not, at first, disturb the idea of natural selection as a kind of beneficent provision of divine government. It merely operated in the old fashion within each episode of the advancing world of life. Buckland, in 1829, for example, speaks of "the carnivora in each period of the world's history fulfilling their destined office—to check excess in the progress of life and maintain the balance of creation." Portlock, a geologist, referred to the idea as late as 1857 as a "sublime conclusion." The conception of a self-regulating balance of nature had survived the discovery of extinction and been taken up in the new progressionism without undergoing the slightest modification. Selection was still a conservative, not a creative, force in nature. Nevertheless, the age of the world had been lengthened and the concept of the struggle for existence was about to enter a new phase.

When Sir Charles Lyell broke through the strata which separated one age from another and proved that all the separate worlds of the catastrophists

were, in reality, one single related and ever altering world, extinction, death, as we have previously
seen, became natural. It was no longer veiled behind the mists of smoking volcanoes or hidden in
the onrush of tidal waves. Species, it became apparent, died as men died, singly and sporadically.
Lyell, like Brückner, continued to see natural
selection as a conservative force. He was faced,
however, with an increasing handicap. If there
was, in actuality, only one ever altering world, then
the balance, the self-regulating government of living things, might not exist. If death approached
living things in this piecemeal fashion, everything
in time might perish.

The only solution lay in the acceptance of the
ideas of the rejected evolutionists or in some form
of mysterious point-creation to replace the sporadic
disappearance of species. Though Lyell wrote
much upon the struggle for existence, in the end
he remained content with its negative aspects. He
saw the web of life as drawn so tightly that there
would be no room for a new form to emerge or
evolve. Before it could do so, he believed, it would
be overwhelmed by the perfectly adjusted organisms about it. At the same time, Lyell was not
insensible to a certain degree of variability within
the limits of a species. It was his opinion that this
ability to develop minor geographic varieties aided
the survival of wide-ranging species. In the main,

however, Lyell insisted upon his principle of pre-occupancy—that is, the idea that plants and animals long adapted to a specific environment will be able to keep foreign intruders from occupying a new country.

Basically this was once more a static conception. At the time it was made, Darwin was observing the influx of foreign weeds in the pampas, the havoc wrought by the introduction of the fauna of the continents upon oceanic islands. But Darwin was still a young naturalist enjoying his first experience of the world. Back home in England, the century was pressing on.

In 1835 Edward Blyth, a young naturalist of Darwin's own age, wrote a paper on animal varieties. Hidden obscurely in the midst of the paper was Blyth's discussion of what he called his "localizing principle." Blyth, in fact, had described what today we call Darwinian natural selection. Blyth saw his principle as one "intended by Providence to keep up the typical qualities of a species." In this respect Blyth sounds like a pure eighteenth-century exponent of natural government, but by 1837 certain additional thoughts had begun to impress themselves upon the young man.

"A variety of important considerations here crowd upon the mind," Blyth confesses, "foremost of which is the enquiry that, as man, by removing species from their appropriate haunts, superinduces changes on their physical constitution and adaptations, to what extent may not the same take place in wild nature, so that, in a few generations, distinctive characters may be acquired, such as are recognized as indicative of specific diversity. *May not then, a large proportion of what are considered species have descended from a common heritage?"*

These words were written in 1837, a short time after Charles Darwin had returned from the *Beagle* voyage. We know that Darwin was an avid reader of the *Magazine of Natural History,* in which Blyth's papers appeared. We know further that there is interior evidence, in Darwin's two early essays before the *Origin of Species* was attempted, which points strongly to Darwin's early knowledge of Blyth's work. At least one of his letters speaks of his fascination with "some few naturalists" in the *Magazine of Natural History,* and there are other references in the undestroyed portions of his note-book recently published by Sir Gavin de Beer. Just as Coleridge had commented eighteen years before, however, scientific convention kept young Edward Blyth from quite accepting his own speculation. "There is a compact among the learned," Coleridge had said, "not to pass beyond a certain limit in speculative science."

Blyth had not made a world voyage like Darwin. He was overawed by his masters, including Lyell. He had rubbed his eyes and the safe and sane conventional world of the English hedgerows had seemed for an instant to alter into something demonic. In the next moment the vision was gone.

Scientific convention has held for a hundred years that the author of the *Origin* read Malthus' *An Essay on the Principle of Population* in 1838 and received from that work the hint which led to his discovery of natural selection. While Malthus undoubtedly had a wide influence during this period and is a convenient source of reference, it is now unlikely that he was Darwin's main source of inspiration. Darwin opened his first notebook on the species question in 1837, the year young Blyth ventured beyond the unconscious convention of his time. Locked away in Darwin's early essays are some curious similarities of thought which strain credulity to be called coincidence.

John Brückner preceded and was used by Thomas Malthus. Edward Blyth preceded and was used by Charles Darwin. There is one difference. Malthus added almost nothing to the thought of Brückner. Darwin, by contrast, took the immature thought, the nudge from Blyth, and combined it with his own vast and growing experience. He refused to see environments as fixed, and organisms as fixed with them. He argued cogently that no environment is completely static, and that selection therefore is constantly at work in the production of new organisms as time and slowly changing geological conditions alter the existing world. An idea which began as an explanation of "natural government," a stabilizing factor making for providential control of the living world, had changed by slow degrees, even as men openly contradicted what their own eyes professed to see.

The individual variation which all organisms revealed, the hereditary alterations produced by the breeder's art, the unreturning fossils in the rocks, were finally combined in the minds of Charles Darwin and Alfred Russel Wallace with the world of competitive struggle, with that concept of plenitude which was now seen to eternally jostle the living in and out of existence.

It was not natural selection that was born in 1859, as the world believes. Instead it was natural selection without balance. Brückner's "impetuous torrent," which he visualized as beating against its safe restraining dikes, is loose and rolling. The violence in Hutton's raindrop is equaled, if not surpassed, by the violence contained in a microscopic genetic particle. The one, multiplied, carries away a mountain range. The other crosses an ice age and produces, on its far side, a man-ape whose intellectual powers now endanger his own civilization.

The world of geological prophecy has vanished. There is only this vast uneasy river of life spreading into every possible niche, dreaming its way toward every possible form. Since the beginning there have been no breaks in that river. The immaterial blueprints were an illusion generated by physical descent. The lime in our bones, the salt in our blood were not from the direct hand of the Craftsman. They were, instead, part of our heritage from an ancient and forgotten sea.

Yet for all this flood of change, movement and destruction, there is an enormous stability about the morphological plans which are built into the great phyla—the major divisions of life. They have all, or most of them, survived since the first fossil records. They do not vanish. The species alter, one might say, but the *Form*, that greater animal which stretches across the millennia, survives. There is a curious comfort in the discovery. In some parts of the world, if one were to go out into the woods, one would find many versions of oneself, with fur and grimaces, surveying one's activities from behind leaves and thickets. It is almost as though somewhere outside, somewhere beyond the illusions, the several might be one.

IV

MANY years ago I was once, by accident, locked in a museum with which I had some association. In the evening twilight I found myself in a lengthy hall containing nothing but Crustacea of all varieties. I used to think they were a rather limited order of life, but as I walked about impatiently in my search for a guard, the sight began to impress, not to say overawe, me.

The last light of sunset, coming through a window, gilded with red a huge Japanese crab on a pedestal at one end of the room. It was one of the stilt-walkers of the nightmare deeps, with a body the size of a human head carried tiptoe on three-foot legs like fire tongs. In the cases beside him there were crabs built and riveted like Sherman tanks, and there were crabs whose claws had been flattened into plates that clapped over their faces and left them shut up inside with little secrets. There were crabs covered with chitinous thorns that would have made them indigestible; there were crabs drawn out and thin, with delicate elongated pinchers like the tools men use to manipulate at a distance in dangerous atomic furnaces.

There were crabs that planted sea growths on their backs and marched about like restless gardens. There were crabs as ragged as waterweed or as smooth as beach pebbles; there were crabs that climbed trees and crabs from beneath the polar ice. But the sea change was on them all. They were one, one great plan that flamed there on its pedestal in the sinister evening light, but they were also many and the touch of Maya, of illusion, lay upon them.

I was shivering a little by the time the guard came to me. Around us in the museum cases was an old pattern, out of the remote sea depths. It was alien to man. I would never underestimate it again. It is not the individual that matters; it is the Plan and the incredible potentialities within it. The forms within the Form are endless and their emergence into time is endless. I leaned there, gazing at that monster from whom the forms seemed flowing, like the last vertebrate on a world whose sun was dying. It was plain that they wanted the planet and meant to have it. One could feel the massed threat of them in this hall.

"It looks alive, Doctor," said the guard at my elbow.

"Davis," I said with relief, "you're a vertebrate. I never appreciated it before, but I do now. You're a vertebrate, and whatever else you are or will be, you'll never be like that thing in there. Never in ten million years. I believe I'm right in congratulating you. Just remember that we're both vertebrates and we've got to stick together. Keep an eye on them now—all of them. I'll spell you in the morning."

Davis did something then that restored my confidence in man. He laughed, and touched my shoulder lightly. I have never heard a crab laugh and I never expect to hear one. It is not in the pattern of the arthropods.

Yet those crabs taught me a lesson really. They reminded me that an order of life is like a diamond of many reflecting surfaces, each with its own pinpoint of light contributing to the total effect. It is a troubling thought, contend some, to be a man and a God-created creature, and at the same time to see animals which mimic our faces in the forest. It is not a good thing to take the center of the stage and to feel at one's back the amused little eyes from the bush. It is not a good thing, someone maintained to me only recently, that animals should stand so close to man.

It depends, I suppose, on the point of view.

On my office wall is a beautiful photograph of a slow loris with round, enormous eyes set in the spectral face of a night-haunter. From a great bundle of fur a small hand protrudes to clasp a branch. Only a specialist would see in that body the far-off simulacrum of our own. Sometimes when I am very tired I can think myself into the picture until I am wrapped securely in a warm coat with a fine black stripe down my spine. And my hands would still grasp a stick as they do today.

At such times a great peace settles on me, and with the office door closed, I can sleep as lemurs sleep tonight, huddled high in the great trees of two continents. Let the storms blow through the streets of cities; the root is safe, the many-faced animal of which we are one flashing and evanescent facet will not pass with us. When the last seared hand has flung the last grenade, an older version of that hand will be stroking a clinging youngster hidden in its fur, high up under some autumn moon. I will think of beginning again, I say to myself then, sleepily. I will think of beginning again, in a different way. . . .

IV

HOW MAN BECAME NATURAL

Here below to live is to change, and to be perfect is to have changed often.

CARDINAL NEWMAN

In my home country, near a small town now almost vanished from the map, there is a region which remains to this day uncultivated. It is best seen at nightfall, because it is then that the full mystery of the place seizes upon the mind. Red granite boulders hundreds of miles from their point of origin thrust awkwardly out of the sparse turf. In the last glow from the west one gets the impression of a waste over which has passed something inhumanly remote and terrifying—something that has happened long ago, but which here lies close to the surface. Crows circle above it like disturbed black memories which rise and fall but never come to rest. It is a barren and disordered landscape, which remembers, and perhaps again anticipates, the cold of glacial ice. It has nothing to do with man; its gravels, its red afterglow, are remnants of another era, in which man was of no consequence.

One instinctively feels that if anything were to graze here it would be mammoth—great sullen, shapeless hulks in the dead light. No one crosses these fields. An invisible barrier confronts one at every turn. Man literally ends here. Beyond lies something morosely violent, of which we have no knowledge, or of which it might be better said that we have a traumatic eagerness to forget. For lurking in this domain is still the nature that created us: the nature of ringing ice fields, of choked forests, of unseasonable thunders. It is the un-predictable nature of the time before the gods, before man had laid hold upon any powers with his mind. It is a season of helplessness that stirs our submerged memories and that causes us to turn back at twilight to the safe road and the lights of town. Behind us whisper the ancient, uncontrollable winds. An animal cries harshly from the dark field we refuse to enter. But enter it we must, though the effort lifts a long-vanished ruff of hair along our nape. This is man's place of birth, this region of inarticulate terror—and this is the story of how he came through the clouds of forget-fulness to find himself in a world which has vanished.

All across Europe, from the British Isles through the North German plain to the Alps and well beyond into Russia, lie the transported boulders of an event so massive in its effects and so strange in its mechanical explanation that European man for a long time failed to digest its meaning. The same marks of that vanished episode—the closest thing to a real cataclysmic event that the planet has ever produced—stretch from sea to sea across temperate America. It is a curious coincidence in history that the moraines of the great continental ice fields should lie across those regions where men believed most implicitly in world deluges and the

iv. *How Man Became Natural*

reality of incalculable violence. The coincidence is not without meaning. Western Christian man, both in America and Europe, saw—where huge boulders lay scattered in windrows—the signs of the passage of a giant tidal wave.

European man was surrounded by sea. He knew its force. In no other manner could he account for those isolated and far-flung monuments of the past. His religious mythology made him even more ready to accept this version of events. "It is interesting to realize," remarked Philip Lake in 1930, on the hundredth anniversary of the publication of Lyell's *Principles of Geology,* "how strong was the case for a great debacle until the glacial deposits of northern Europe had been recognized as glacial." Ironically enough, man had begun to peer into the dark field of his origin, in just that region which offered the greatest hints of disturbance, of mystery, of an inviolate line across which it was impossible to pass.

"If species have changed by degrees," wrote Cuvier in 1815, "we ought to find traces of this gradual modification. We should be able to discover some intermediate forms; and yet no discovery has ever been made." After reviewing the then existing knowledge of fossil elephants and other creatures nearly allied to those of the present, Cuvier says of the coming of man that it "must necessarily have been posterior . . . to the revolutions which covered up these bones. No argument for the antiquity of the human race in those countries can be founded upon these fossil bones or upon the rocks by which they are covered." Humanity, Cuvier asserts, "did not exist in the countries in which the fossil bones of animals have been discovered."

In Cuvier's time, and for many years thereafter, during the reign of catastrophic geology, man's thoughts about his past would falter inevitably before those boulder-strewn wastes in which the human story seemed to end abruptly. Although Cuvier's single reservation was lost in later, more enthusiastically religious volumes, it can be noted that the master anatomist did have a moment of hesitation. "I do not presume," he said, "to conclude that man did not exist at all before these epochs. He may have then inhabited some narrow region, whence he went forth to re-people the earth after the cessation of these terrible revolutions. Perhaps even the places which he then inhabited may have sunk into the abyss."

As the pre-Darwinian years began to pass, men returned again and again to the contemplation of what seemed the shallow time-depths of the human species in Europe. Industrial activities here and there resulted in the disturbance of ancient strata. A few bones and crude implements faintly suggested that beneath historic Europe, the Europe of the Romans and Greeks, might lie an unknown, shadowy world in which the white race had cracked marrow bones by campfires as primitive as those of red Indians. It was not a nice thought to intrude into civilized minds. Most people rejected it. "Man," insisted Darwin's geology teacher, Sedgwick, "has been but a few years' dweller on the earth. He was called into being within a few thousand years of the days in which we live by a provident contriving power."

If any association of the bones of Cuvier's mammoths with men was brought forward, it was suggested that the human remains and the fossils had been accidentally mixed in later times. The first paleolithic archaeologists were apt to find themselves doubted by their more conservative colleagues. No one had ever seen such crude tools before. Were they really of human manufacture? Even Darwin, in his earlier years, had been dubious. To make things worse, workmen faked finds to gain shillings or francs from overeager investigators. Thus fraud confused the issue.

But the boulders still stretched in mysterious lines across Europe. Hutton's inspired guess about ice action had long since been forgotten. Even his intellectual descendant, Charles Lyell, speculated that the stones were transported by icebergs in greater seas. As Sir James Geikie, the great Scotch geologist, was later to observe, "Even a cautious thinker like Lyell saw less difficulty in sinking the whole of Central Europe under the sea and covering the waters with floating icebergs, than in conceiving that the Swiss glaciers were once large enough to reach to the Jura. Men shut their eyes to the meaning of the unquestionable fact that, while there was absolutely no evidence for a marine submergence, the former track of the glaciers could be followed mile after mile . . ." The episode is again an apt illustration of how difficult it is even for trained scientists to break out of a prevailing mode of thought.

II

IN 1837, however, two remarkable discoveries occurred. They did not change the intellectual climate of the age overnight. Nevertheless they resulted, eventually, in a drastic revision of the ruling conception of human antiquity and of the nature of those boulder-strewn fields whose wastes had long tantalized and intimidated the inquiring human spirit.

The first of these events was the discovery of the fossil remains of great apes and other primates in the Tertiary beds of northern India. It had long been assumed that man's nearest anatomical relatives were totally contemporaneous with himself, and were, in catastrophist terms, part of the living creation which included man.

In 1830, Hugh Falconer, a young Scotch doctor with a deep interest in natural history, had gone out to India as an assistant surgeon in the service of the East India Company. He rapidly acquired a reputation as a scientist and in 1832 became superintendent of the Botanic Gardens at Suharunpoor, which lies about twenty-five miles from the Siwalik hills, an eroded Tertiary outlyer of the Himalayas. Here Falconer, with the assistance of army friends, began the paleontological explorations which were to lead to the discovery of one of the finest Tertiary fossil beds in the world. Falconer, with his friend Captain Proby Cautley, established the age of the deposits and brought to light a subtropical mammalian fauna which excelled in richness any other Cenozoic fossil beds then known.

Although Cautley and Falconer made the first discovery of a fossil primate from these beds, they deferred publication because of the fragmentary nature of their discovery. In the meantime, their friends W. E. Baker and H. M. Durand, of the army engineers, located and reported upon a much larger species, equivalent in size to the orang. The specimen, they indicated, revealed "the existence

of a gigantic species of quadrumanous animal con-
temporaneously with the Pachydermata of the
sub-Himalayas and thus supplies . . . proof of the
existence, in a fossil state, of the type of organization most nearly resembling that of man."

Here at last was more than a new fossil for the
taxonomist. Here, in Falconer's words, was a mix-
ture of the old and new together, "affording another illustration of constancy in the order of
nature, of an identity of condition in the earth with
what it exhibits now." Most scientists were not
yet evolutionists, but the barrier to the past which
they had encountered in Europe had been most
dramatically pierced on the far-off flanks of the
Himalayas. There "was clear evidence, physical
and organic," Falconer reminisced later, "that the
present order of things had set in from a very
remote period in India. Every condition was suited
to the requirements of man. The wide spread of
the plains of India showed no signs of the un-
stratified superficial Gravels, Sands and Clays,
which for a long time were confidently adduced as
evidence that a great Diluvial wave had suddenly
passed over Europe and other continents, over-
whelming terrestrial life."

It was not in Europe, nor in North America, that
the earliest relics of the human race were to be
sought. Man in those regions, argued Falconer,

was a creature of yesterday. It was in the great alluvial valleys of the tropics and subtropics that his earlier history would be found.

The second discovery which was drastically to alter our notions of the environment of prehistoric man was made, or rather reintroduced into science, in the same year that Falconer and his friends were busying themselves with the Siwalik primates. Louis Agassiz, whose remarkable researches upon the fossil fishes of Europe had brought him international acclaim, turned, at the age of thirty-three, to a study of glaciation. Before his time, save for the premonitory glimpses of a few men, including Hutton, who did not pursue the problem, the European and American ice erratics were accounted for in one of two ways: as the product of turbulent flood waters of oceanic proportions, or as the result of transportal by floating ice drifting over submerged areas.

Agassiz studied the effects of ice action in modern glaciers and observed the marks of scouring and engraving which no water action could duplicate. He was the first to recognize the enormous extent of the continental ice fields in both Europe and America. He grasped equally well the fact that such a widespread phenomenon must involve causes at work throughout the whole northern hemisphere.

The British, in particular, were loath to give up the conception of sea-borne ice, but in the end Agassiz carried the day and became, in the process, the founder of glacial geology. Flood-swept Europe became ice-locked Europe, though, as we have intimated, it was growing apparent that man—at least late glacial man—had managed to subsist under rude conditions in caves and rock shelters south of the ice desert. There had been no diluvial wave to carry him off.

The crude tools associated with the bones of extinct animals, which had been so long scorned, were read. There actually *was* a way through the doorway of the past. Paradoxically, Agassiz lived on to oppose Darwin and reiterate his belief in world-wide extinctions. Falconer, too, though more temperate and a warm friend of Darwin's, had reservations about the reality of evolution.

Nevertheless, the two men had opened new vistas. We have already examined the mechanism which Darwin had provided as a natural explanation of organic change. We will now want to observe the manner in which the Darwinian circle applied this conception to human evolution. The creation of natural man encountered unexpected obstacles, not all of which were provided by resentful theologians. Darwin and his followers, carrying over concepts which had proved useful in the other realms of biology, were not always judicious in

their examination of man. Their biological tri-
umph had been so complete, however, that they,
in turn, had created an intellectual climate which
accepted their views unquestioningly.

III

WE HAVE already had occasion to observe that to
the catastrophist school of thought man was not
the incidental product of variation in a ground-
dwelling ape, but rather a creature foreordained
and foreseen from the beginning of geologic time.
When evolution began to be taken seriously in
England, and it seemed inevitable that it would be
extended to man, the latter's peculiarly human
characteristics began to be scrutinized with care.
The biologist sought to "animalize" man, even if
he had to humanize the animal world in order to
demonstrate a genetic connection.

The advocate of man's divinity, on the other
hand, sought to identify human traits of which
there was no trace in the surrounding organic
world and thus to confront the Darwinists with a
break, or discontinuity, in their system. This was
apt to be particularly nettling, because Darwin had
placed such emphasis upon slow, almost invisible
change over vast time periods. It needs scarcely to
be said that both sides were forced to argue almost

entirely on theoretical grounds, for although human antiquity had been greatly extended after the full understanding of the ice age, only one extinct specimen of man had been doubtfully recognized, while none of the really subhuman links in the phylogeny of man were yet known. Human evolution, therefore, was for long debated upon largely abstract grounds. Indeed, the very lack of human fossils was occasionally pointed out as an argument against attaching man to the rest of the organic world. To many thinkers he remained a divine interposition in the universe.

I have said that the biologist of that day, lacking enough fossils to substantiate his points, sought to thrust living men and living animals closer together than the facts sometimes warranted. This was not a conscious attempt at deception, but again was the product of a particular intellectual climate. There was, for example, a tendency to see human evolution in terms of the eighteenth-century scale of nature as based on living forms. Thus one could begin with one of the existing great apes, the gorilla, orang or chimpanzee, as representing the earliest "human" stage, and from this pass to the existing human races, which were arranged in a vertical series with white Victorian man representing the summit of evolutionary ascent.

The fact that the existing apes are creatures

contemporaneous with ourselves, who have evolved down another evolutionary pathway which is also highly specialized, was lost from sight. Even the contemporary races of man came to be regarded as living fossils. Today this whole approach to the human problem has been abandoned, but it was peculiarly attractive to many nineteenth-century thinkers. The past, so to speak, had been quietly transported into the present, and evolutionary roles—not always attractive ones—assigned to living actors without their consent. As a matter of fact, a curious twofold interpretation of the human psyche has descended from the Darwinian epoch into modern science. Darwin himself seems to have hesitated in his book, *The Descent of Man,* between a conception of man the warrior—the product of ruthless, competitive forces—and man the weak-bodied, unarmed primate who, until his intellectual powers were strengthened, might not have been able to survive on the ground in competition with the great carnivores.

Today it is frequently recalled, because of fossil discoveries made there in the last decade, that Darwin once suggested Africa as the possible original home of man. What is less often remembered is the fact that Darwin, in the same volume, suggested that man, even if far more helpless and defenseless than any existing savages, would not have been exposed to any special danger had he inhabited some warm continent or large island such as Australia, New Guinea or Borneo. The creation of a sort of Garden of Eden for early man seems to have been forced upon Darwin by the criticism of the Duke of Argyll, who pointed out that physically man is one of the most helpless creatures in the world.

The Duke raised the question, in the light of Darwin's and Huxley's emphasis on the struggle for existence, how, by their own philosophy, the human body could have diverged in the direction of greater helplessness in the early phase of its evolution. This criticism forced upon Darwin the wary evasion of his island Eden. "We do not know," he admitted, "whether man is descended from some small species, like the chimpanzee, or from one as powerful as the gorilla. Therefore we cannot say whether man has become larger and stronger or smaller and weaker than his ancestors." He confessed, however, that an animal as powerful as the gorilla might not have become social.

In spite of this hesitation as to the precise nature of early man, it is evident that the existing great apes played a large part in Darwinian thinking and that some gorilloid characters came to be projected upon the first Neanderthal fossils. Such characters seemed appropriate in the light of the Darwinian interpretation of the struggle for existence.

IV

Just a little over a hundred years ago two German quarrymen were digging in a small cave along the gorge of the Neander near Düsseldorf, Germany. In a dark interior chamber which was filled to a depth of four or five feet by earth which had been swept into the cave at some time in the long past, the workmen stumbled upon some strange bones. As the deposit of clay and stones was removed, the men came first upon a skull lying near the entrance of the grotto. Later, stretched along the floor of the cave, other skeletal remains appeared. With no idea that they had unearthed a human skeleton destined to become the scientific sensation of its time, these rough and unskilled quarrymen picked up some of the larger bones and tossed them out with the debris of their excavation.

Weeks later, what could be found of the broken and dispersed skeleton was placed in the hands of a local physician, Dr. Fuhlrott of Elberfeld. This individual, an enlightened collector and one of those early scientific pioneers of whom the medical profession affords so many examples, immediately dispatched the bones to a skilled anatomist, Professor Hermann Schaafhausen of Bonn University. Thus, three years before Charles Darwin gave to the world his theory of evolution, science found placed in its hands an important clue to the prehistoric past of the human race.

Today we know that this low-browed, thick-walled skull vault, which its describer promptly characterized as "due to a natural conformation *hitherto not known to exist even in the most barbarous races,"* belongs to an era far more remote than even Schaafhausen dreamed. This skull is man's first relic of the ice age—of a long-vanished world where men and women had endured glacial cold and had struggled for existence armed only with crude spears and sharpened flints.

When this dead man had been interred in the little cave by the gorge of the Neander, all the enormous triumphs of humanity, the piled wealth of our great cities, still lay tens of thousands of years in the future. On the glacial uplands of Europe, winds had howled and dust had swirled endlessly over naked grasslands. Only in sheltered gullies and occasional caves flickered the fires of the sparse human population, which lived by hunting mammoth through the phantom streets of future Paris and Berlin.

By Darwinian standards, these creatures were an odd and unimagined link with the past. Their skulls, in spite of jutting brow ridges and massive chinless faces, had brains as large as or larger than our own. Huxley, the swashbuckling evolutionist, hesitated over their meaning with the reluctance of a choirboy. Darwin saw them as armed with gorilloid fangs, and an artist pictured them with the grasping feet of apes. A distinguished anatomist spoke of them as "the quintessence of brute-be-nightedness."

These were the men who had scouted for game within a few miles of that great blue barrier which had once lain like a sea over all of northern Europe. These were the men whose skin and eye color we do not know, but who fought a battle for naked survival against cold and the prowling beasts thrust southward by the long inexorable advance of the continental ice field. Generation by generation life fell back before it. Ice grumbled across the North German plain; in the Alps the glaciers of the high altitudes moved downward, coalesced and blocked Italy away from the north. The vast northern sheet reduced the corridor between eastern Europe and the west to a narrow and freezing channel between two great ice masses.

Here then was the "ice island." On the south lay the barrier of the Mediterranean Sea; on the north was a blue-white silence which could be matched today only in the Antarctic; a place of winter unbroken save by the crack of ice and the fall of avalanches. Almost where the borders of the Iron Curtain run today, ran the ice curtain. Below it men moved, tiny and overawed, surrounded and cut off from Asia and the world farther east. And still the ice came on, and still, warmed by the Atlantic waters and the Mediterranean, the little groups of naked hunters clung to life.

Season followed season. Men grubbed for roots and nuts and berries in the short, cool summers, and followed remorselessly upon the trail of animals heavy with young. The spear and cutting flints were their only weapons; the bow was as far away in the future as gunpowder from the bow. All killing was personal. If a thrust failed, men stood to the charge of big game with nothing but a wooden spear and a flint knife. In the camps were children to be fed—big-browed children, the sides of whose foreheads were beginning to roof out with the heavy bone of their strange fathers.

Somewhere, perhaps around the forty-thousand-year mark, the ice hesitated. A little more and the men of West Europe might have vanished. They huddled in the caves of France and along the warm Mediterranean shore, little bands composed of a few score people, tough and enduring. Illiterate, they had no knowledge of any other world. This was their life, the sound of ice in the mountains, the endless search for game.

Thousands of years had passed while the ice ground southward, thousands more while the great sheet thrust out exploring tongues into the western valleys, paused, and slowly, very slowly, began a slight retreat. All this time western man in scant numbers had lived alone and without contact. We have information which suggests that outside of Europe a branch of the old big-browed strain was modifying in the direction of today's human type. Here in West Europe, however, the caves of this period yield up only the massive-muzzled creature who seems, in the eyes of modern science, to be in some degree the creation of the ice.

We know two things. We know that wherever small populations of any type of living thing are marooned as Neanderthal man was marooned, a process of genetic drift sets in. That is, mutations, little changes in the germ plasm, may run with comparative quickness over a whole group. If the new physical characters promote survival, so much the better. In any case, little groups of primitive men are always intimately related. The novelist Thomas Hardy expressed this in his poem about family likeness "leaping from place to place over oblivion." Even today, in a few regions in the mountain South, one can observe a physical likeness which is the product of local isolation over several generations. The people are interrelated; a local type has arisen and is revealed in a vague physical similarity. In West Europe such isolation continued for thousands of years. In addition, it may well be that something in that desperate struggle with the ice enhanced the value of these primitive characters.

Now the classic western Neanderthal type has been carefully analyzed. Such notions as those of the Darwinian years that our cave men had threatening gorilloid fangs or clasping feet have long since been abandoned. Nevertheless, these people who were living in the earlier part of the last ice advance in caves in France, Italy, Spain and Belgium are distinctly different from ourselves. Their skulls, which housed a brain in some instances larger than the modern average, were low, long and broad. Where the modern skull vault is high, Neanderthal man, by contrast, had a low but very wide skull. The eye orbits were large and overhung by a pronounced bar of bone suggestive of that which can be observed in a modern chimpanzee or gorilla. The face is massive, the jaw lacking the pointed chin of today.

The chest was barrel-like and the stature short—just slightly over five feet. The forearm was short and powerful, as was the lower leg. The over-all picture is that of a very powerful but economically built man—a creature selected for survival under conditions demanding great hardihood and physical strength sustained often on a minimum of food.

Nevertheless, the Neanderthal people reveal no such quick descent to beastdom as the nineteenth-century writers presumed with their lurid pictures of a creature "in the highest degree hideous and ferocious." Indeed, we have clear evidence that they buried their dead with offerings. There is evidence also that these men were capable of altruistic care of adults. The big-brained primitives who seem to characterize the upper reaches of the ice age in the end forced scholars to reassess the time involved in the human transformation and to extend it. It has become more and more evident with the passing years that the place of the human emergence, whether it be Darwin's fanged gorilla of unsocial nature or his Eden-like creature of comparative helplessness, lies much farther back in the time scale than the nineteenth century realized. The nature of the original animal-man is still a matter of some debate.

V

THE likelihood of man's origin on some idyllic Bornean island has been abandoned long since. However man managed his transformation, it was achieved amidst the great mammals of the Old World land mass, and most probably in Africa. Here the discoveries of the last few decades have revealed small-bodied, upright-walking man-apes whose brains appear little, if at all, larger than those of the existing great apes. Whether they could speak is not known, but it now appears that

they were capable of at least some crude tool-mak-
ing capacities.

As might be expected, these bipedal apes are
quite different from our arboreal cousins, the great
apes of today. Their teeth are heavy-molared, but
they lack completely the huge canine teeth with
which the nineteenth century, using gorilla models,
endowed our ancestors. In spite of a powerful jaw
musculature the creatures are light-bodied and apparently pygmoid, compared with modern man.

In spite, also, of exaggerated guesses when the
first massive jaws were recovered, the several forms
of these creatures in no case indicate man's descent
from a giant primate—quite the reverse, in fact. Instead, we seem to be confronted with a short-faced,
big-jawed ape of quite moderate body dimensions
and a brain qualitatively, if not quantitatively, su-
perior to that of any existing anthropoid.

It is unlikely that all of the several species known
became men. Some, indeed, are late enough in
time that primitive men were probably already in
existence. Surveyed as a whole, however, they suggest an early half-human or near-human level, re-
vealing clearly that the creature who became man
was a ground-dwelling, well-adapted ape long before his skull and brain underwent their final transformation.

Interestingly, the hesitation exhibited by Darwin
over the psychical nature of the human forerunner

continues into these remote time depths of the lower ice age. Raymond Dart, one of the pioneer discoverers of these ape-men, regards them as successful carnivores and killers of big game. He sees them as brutal and aggressive primates, capable of killing their own offspring, and carrying in their genetic structure much of the sadism and cruelty still manifested by modern man. They are club-swingers par excellence, already ably balanced on their two feet, and the terror of everything around them.

One cannot help but feel, however, that Dart tends to minimize the social nature which even early man must have had to survive and care for young who needed ever more time for adaptation to group life. He paints a picture too starkly overshadowed by struggle to be quite believable. It would seem that into his paleontological studies has crept a touch of disillusionment and distaste which has been projected backward upon that wild era in which the human predicament began. This judgment is, of course, a subjective one. It reveals that the hesitations of Darwin's day, in spite of increased knowledge, follow man back into the past. Perhaps it is a matter of temperament on the part of the observer. Perhaps man has always been both saint and sinner—even in his raw beginnings.

As we press farther back in time, however, back until the long, desolate years of the fourth ice lie somewhere far in the remote future, we come upon something almost unbelievable. We come upon man, near-man, "the bridge to man," estimated as some two million years removed from us, in the Olduvai Gorge in Tanganyika. He is a step beyond Dart's man-apes, according to Dr. L. S. B. Leakey, his discoverer. Like them, he is small, huge-jawed, with a sagittal crest like that of a gorilla, but a true shaper of tools. Unlike Dart, Leakey describes his specimen as a semivegetarian eater of nuts, small rodents and lizards. Perhaps massive jaws and molar teeth, just as in the case of the living gorilla, were developed for other food than flesh.

The thing which appears the strangest of any news to come down from that far epoch, however, is the report from London that the youthful ape-man's body had apparently been carefully protected from scavenging hyenas until rising lake waters buried him among his tools.

Three years ago, in a symposium on the one hundredth anniversary of the discovery of Neanderthal man, I made these remarks: "When we consider this creature of 'brute benightedness' and 'gorilloid ferocity,' as most of those who peered into that dark skull vault chose to interpret what they saw there, let us remember what was finally revealed at the little French cave near La Chapelle-aux-Saints in 1908. Here, across millennia of time, we can observe a very moving spectacle. For these men whose brains were locked in a skull foreshadowing the ape, these men whom scientists had contended to possess no thoughts beyond those of the brute, had laid down their dead in grief.

"Massive flint-hardened hands had shaped a sepulcher and placed flat stones to guard the dead man's head. A haunch of meat had been left to aid the dead man's journey. Worked flints, a little treasure of the human dawn, had been poured lovingly into the grave. And down the untold centuries the message had come without words: 'We too were human, we too suffered, we too believed that the grave is not the end. We too, whose faces affright you now, knew human agony and human love.'

"It is important to consider," I said then, "that across fifty thousand years nothing has changed or altered in that act. It is the human gesture by which we know a man, though he looks out upon us under a brow reminiscent of the ape."

If the London story is correct, an aspect of that act has now been made distant from us by almost a million years. The creature who made it could only be identified by specialists as human. He is far more distant from Neanderthal man than the latter is from us.

Man, bone by bone, flint by flint, has been traced backward into the night of time more successfully than even Darwin dreamed. He has been traced to a creature with an almost gorilloid head on the light, fast body of a still completely upright, plains-dwelling creature. In the end he partakes both of Darwinian toughness, resilience, and something else, a humanity—if this story is true—that runs well nigh as deep as time itself.

Man has, in scientific terms, become natural, but the nature of his "naturalness" escapes him. Perhaps his human freedom has left him the difficult choice of determining what it is in his nature to be. Perhaps the two sides of the dark question Darwin speculated upon were only an evolutionary version of man's ancient warfare with himself—a drama as great in its hidden fashion as the story of the Garden and the Fall.

V

HOW HUMAN IS MAN?

---

*Be not under any Brutal Metemp-
sychosis while thou livest and walkest
about erectly under the scheme
of Man.*

SIR THOMAS BROWNE

Over a hundred years ago a Scandinavian philosopher, Sören Kierkegaard, made a profound observation about the future. Kierkegaard's remark is of such great, though hidden, importance to our subject that I shall begin by quoting his words. "He who fights the future," remarked the philosopher, "has a dangerous enemy. The future is not, it borrows its strength from the man himself, and when it has tricked him out of this, then it appears outside of him as the enemy he must meet."

We in the western world have rushed eagerly to embrace the future—and in so doing we have provided that future with a strength it has derived from us and our endeavors. Now, stunned, puzzled and dismayed, we try to withdraw from the embrace, not of a necessary tomorrow, but of that future which we have invited and of which, at last, we have grown perceptibly afraid. In a sudden horror we discover that the years now rushing upon us have drained our moral resources and have taken shape out of our own impotence. At this moment, if we possess even a modicum of reflective insight, we will give heed to Kierkegaard's concluding wisdom: "Through the eternal," he enjoins us, "we can conquer the future."

The advice is cryptic; the hour late. Moreover,

what have we to do with the eternal? Our age, we know, is littered with the wrecks of war, of outworn philosophies, of broken faiths. We profess little but the new and study only change.

Three hundred years have passed since Galileo, with the telescope, opened the enormous vista of the night. In those three centuries the phenomenal world, previously explored with the unaided senses, has undergone tremendous alteration in our minds. A misty light so remote as to be scarcely sensed by the unaided eye has become a galaxy. Under the microscope the previously unseen has become a cosmos of both beautiful and repugnant life, while the tissues of the body have been re-solved into a cellular hierarchy whose constituents mysteriously produce the human personality.

Similarly, the time dimension, by the use of other sensory extensions and the close calculations made possible by our improved knowledge of the elements, has been plumbed as never before, and even its dead, forgotten life has been made to yield remarkable secrets. The great stage, in other words, the world stage where the Elizabethans saw us strutting and mouthing our parts, has the skele-tons of dead actors under the floor boards, and the dusty scenery of forgotten dramas lies abandoned in the wings. The idea necessarily comes home to us then with a sudden chill: What if we are not playing on the center stage? What if the Great

Spectacle has no terminus and no meaning? What if there is no audience beyond the footlights, and the play, in spite of bold villains and posturing heroes, is a shabby repeat performance in an echoing vacuity? Man is a perceptive animal. He hates above all else to appear ridiculous. His explorations of reality in the course of just three hundred years have so enlarged his vision and reduced his ego that his tongue sometimes fumbles for the proper lines to speak, and he plays his part uncertainly, with one dubious eye cast upon the dark beyond the stage lights. He is beginning to feel alone and to hear nothing but echoes reverberating back.

It will do no harm then, if in this moment of hesitation we survey the history of our dilemma. Man's efforts to understand his predicament can be compassed in the simple mechanics of the theatre. We have examined the time allowed the play, the nature of the stage, and what appears to be the nature of the plot. All else is purely incidental to this drama, and it may well be that we can see our history in no other terms, being mentally structured to look within as well as without, and to be influenced within by what we consider the nature of the "without" to be. It is for this reason that the "without," and our modes of apprehending it, assume so pressing an importance. Nor is it fully possible to understand the human drama, the drama of the great stage, without a historical knowledge of how the characters have interpreted their parts in the play, and in doing so perhaps affected the nature of the plot itself.

This, in brief, epitomizes the role of the human mind in history. It has looked through many spectacles in the last several centuries, and each time the world has appeared real, and the plot has been played accordingly. Strange colorings have been given to reality and the colors have come mostly from within. As science extends itself, the colors, and through them the nature of reality, continue to change. The "within" and "without" are in some strange fashion intermingled. Perhaps, in a sense, the great play is actually a great magic, and we, the players, are a part of the illusion, making and transforming the plot as we go.

If the play has its magical aspect, however, there is an increasing malignancy about it. A great Russian novelist ventures to remark mildly that the human heart, rather than the state, is the final abode of goodness. He is immediately denounced by his colleagues as a heretic. In the West, psychological studies are made of human "rigidity," and although there is a dispassionate scientific air about them the suggestion lingers that the "normal" man should conform; that the deviant is pathological. The television networks seek the lowest denominator which will entrance their mass audience. There is a muted intimation that we can do without the

kind of intellectual individualists who used to de-
claim along the edges of the American wilderness
and who have left the world some highly explosive
literature in the shape of *Walden* and *Moby Dick*.
It is obvious that the whole of western ethic,
whether Russian or American, is undergoing
change, and that the change is increasingly toward
conformity in exterior observance and, at the same
time, toward confusion and uncertainty in deep
personal relations. In our examination of this
phenomenon there will emerge for us the meaning
of Kierkegaard's faith in the eternal as the only
way of achieving victory against the corrosive
power of the human future.

II

IF WE examine the living universe around us which
was before man and may be after him, we find two
ways in which that universe, with its inhabitants,
differs from the world of man: first, it is essentially
a stable universe; second, its inhabitants are in-
tensely concentrated upon their environment.
They respond to it totally, and without it, or rather
when they relax from it, they merely sleep. They
reflect their environment but they do not alter it.
In Browning's words, "It has them, not they it."
Life, as manifested through its instincts, de-

mands a security guarantee from nature that is largely forthcoming. All the release mechanisms, the instinctive shorthand methods by which nature provides for organisms too simple to comprehend their environment, are based upon this guarantee. The inorganic world could, and does, exist in a kind of chaos, but before life can peep forth, even as a flower, or a stick insect, or a beetle, it has to have some kind of unofficial assurance of nature's stability, just as we have read that stability of forces in the ripples impressed in stone, or the rain marks on a long-vanished beach, or the unchanging laws of light in the eye of a four-hundred-million-year-old trilobite.

The nineteenth century was amazed when it discovered these things, but wasps and migratory birds were not. They had an old contract, an old promise, never broken till man began to interfere with things, that nature, in degree, is steadfast and continuous. Her laws do not deviate, nor the seasons come and go too violently. There is change, but throughout the past life alters with the slow pace of geological epochs. Calcium, iron, phosphorus, could exist in the jumbled world of the inorganic without the certainties that life demands. Taken up into a living system, however, *being* that system, they must, in a sense, have knowledge of the future. Tomorrow's rain may be important, or tomorrow's wind or sun. Life, in contrast to the inorganic, is historic in a new way. It reflects the past, but must also expect something of the future. It has nature's promise—a guarantee that has not been broken in three billion years—that the universe has this queer rationality and "expectedness" about it. "Whatever interrupts the even flow and luxurious monotony of organic life," wrote Santayana, "is odious to the primeval animal."

This is a true observation, because on the more simple levels of life, monotony is a necessity for survival. The life in pond and thicket is not equipped for the storms that shake the human world. Its small domain is frequently confined to a splinter of sunlight, or the hole under a root. What life does under such circumstances, how it meets the precarious future (for even here the future can be precarious), is written into its substance by the obscure mechanisms of nature. The snail recoils into his house, the dissembling caterpillar who does not know he dissembles, thrusts stiffly, like a budding twig, from his branch. The enemy is known, the contingency prepared for. But still the dreaming comes from below, from somewhere in the molecular substance. It is as if nature in a thousand forms played games against herself, but the games were each one known, the rules ancient and observed.

It is with the coming of man that a vast hole seems to open in nature, a vast black whirlpool spinning faster and faster, consuming flesh, stones, soil, minerals, sucking down the lightning, wrenching power from the atom, until the ancient sounds of nature are drowned in the cacophony of something which is no longer nature, something instead which is loose and knocking at the world's heart, something demonic and no longer planned—escaped, it may be—spewed out of nature, contending in a final giant's game against its master.

Yet the coming of man was quiet enough. Even after he arrived, even after his strange retarded youth had given him the brain which opened up to him the dimensions of time and space, he walked softly. If, as was true, he had sloughed instinct away for a new interior world of consciousness, he did something which at the same time revealed his continued need for the stability which had preserved his ancestors. Scarcely had he stepped across the border of the old instinctive world when he began to create the world of custom. He was using reason, his new attribute, to remake, in another fashion, a substitute for the lost instinctive world of nature. He was, in fact, creating another nature, a new source of stability for his conflicting erratic reason. Custom became fixed: order, the new order imposed by cultural discipline, became the "nature" of human society. Custom directed the vagaries of the will. Among the fixed institutional bonds of society man found once more the security of the animal. He moved in a patient renewed orbit with the seasons. His life was directed, the gods had ordained it so. In some parts of the world this long twilight, half in and half out of nature, has persisted into the present. Viewed over a wide domain of history this cultural edifice, though somewhat less stable than the natural world, has yet appeared a fair substitute—a structure, like nature, reasonably secure. But the security in the end was to prove an illusion. It was in the West that the whirlpool began to spin. Ironically, it began in the search for the earthly Paradise.

The medieval world was limited in time. It was a stage upon which the great drama of the human Fall and Redemption was being played out. Since the position in time of the medieval culture fell late in this drama, man's gaze was not centered scientifically upon the events of an earth destined soon to vanish. The ranks of society, even objects themselves, were Platonic reflections from eternity. They were as unalterable as the divine Empyrean world behind them. Life was directed and fixed from above. So far as the Christian world of the West was concerned, man was locked in an unchanging social structure well nigh as firm as nature. The earth was the center of divine attention. The ingenuity of intellectual men was turned almost exclusively upon theological problems.

As the medieval culture began to wane toward

its close, men turned their curiosity upon the world around them. The era of the great voyages, of the breaking through barriers, had begun. Indeed, there is evidence that among the motivations of those same voyagers, dreams of the recovery of the earthly Paradise were legion. The legendary Garden of Eden was thought to be still in existence. There were stories that in this or that far land, behind cloud banks or over mountains, the abandoned Garden still survived. There were speculations that through one of those four great rivers which were supposed to flow from the Garden, the way back might still be found. Perhaps the angel with the sword might still be waiting at the weed-grown gateway, warning men away; nevertheless, the idea of that haven lingered wistfully in the minds of captains in whom the beliefs of the Middle Ages had not quite perished.

There was, however, another, a more symbolic road into the Garden. It was first glimpsed and the way to its discovery charted by Francis Bacon. With that act, though he did not intend it to be so, the philosopher had opened the doorway of the modern world. The paradise he sought, the dreams he dreamed, are now intermingled with the countdown on the latest model of the ICBM, or the radioactive cloud drifting downwind from a megaton explosion. Three centuries earlier, however, science had been Lord Bacon's road to the earthly

Paradise. "Surely," he wrote in the *Novum Organum,* "it would be disgraceful if, while the regions of the material globe, that is, of the earth, of the sea, and of the stars—have been in our times laid widely open and revealed, the intellectual globe should remain shut up within the narrow limits of the old discoveries."

Instead, Bacon chafed for another world than that of the restless voyagers. "I am now therefore to speak touching Hope," he rallied his audience, who believed, many of them, in a declining and decaying world. Much, if not all, that man lost in his ejection from the earthly Paradise might, Bacon thought, be regained by application, so long as the human intellect remained unimpaired. "Trial should be made," he contends in one famous passage, "whether the commerce between the mind of men and the nature of things . . . might by any means be restored to its perfect and original condition, or if that may not be, yet reduced to a better condition than that in which it now is." To the task of raising up the new science he devoted himself as the bell ringer who "called the wits together."

Bacon was not blind to the dangers in his new philosophy. "Through the premature hurry of the understanding," he cautioned, "great dangers may be apprehended . . . against which we ought even now to prepare." Out of the same fountain, he

saw clearly, could pour the instruments of benefi-
cence or death.

Bacon's warning went unheeded. The struggle
between those forces he envisaged continues into
the modern world. We have now reached the point
where we must look deep into the whirlpool of the
modern age. Whirlpool or flight, as Max Picard has
called it, it is all one. The stability of nature on the
planet—that old and simple promise to the living,
which is written in every sedimentary rock—is
threatened by nature's own product, man.

Not long ago a young man—I hope not a forerunner of the coming race on the planet—remarked to me with the colossal insensitivity of the
new asphalt animal, "Why can't we just eventually
kill off everything and live here by ourselves with
more room? We'll be able to synthesize food pretty
soon." It was his solution to the problem of over-
population.

I had no response to make, for I saw suddenly
that this man was in the world of the flight. For
him there was no eternal, nature did not exist save
as something to be crushed, and that second order
of stability, the cultural world, was, for him, also
ceasing to exist. If he meant what he said, pity had
vanished, life was not sacred, and custom was a
purely useless impediment from the past. There
floated into my mind the penetrating statement of
a modern critic and novelist, Wright Morris. "It is

not fear of the bomb that paralyzes us," he writes, "not fear that man has no future. Rather, it is the *nature* of the future, not its extinction, that produces such foreboding in the artist. It is a numbing apprehension that such future as man has may dispense with art, with man as we now know him, and such as art has made him. The survival of men who are strangers to the nature of this conception is a more appalling thought than the extinction of the species."

There before me stood the new race in embryo. It was I who fled. There was no means of communication sufficient to call across the roaring cataract that lay between us, and down which this youth was already figuratively passing toward some doom I did not wish to see. Man's second rock of certitude, his cultural world, that had gotten him out of bed in the morning for many thousand years, that had taught him manners, how to love, and to see beauty, and how, when the time came, to die— this cultural world was now dissolving even as it grew. The roar of jet aircraft, the ugly ostentation of badly designed automobiles, the clatter of the supermarkets could not lend stability nor reality to the world we face.

Before us is Bacon's road to Paradise after three hundred years. In the medieval world, man had felt God both as exterior lord above the stars, and as immanent in the human heart. He was both outside and within, the true hound of Heaven. All this alters as we enter modern times. Bacon's world to explore opens to infinity, but it is the world of the outside. Man's whole attention is shifted outward. Even if he looks within, it is largely with the eye of science, to examine what can be learned of the personality, or what excuses for its behavior can be found in the darker, ill-lit caverns of the brain.

The western scientific achievement, great though it is, has not concerned itself enough with the creation of better human beings, nor with self-discipline. It has concentrated instead upon things, and assumed that the good life would follow. Therefore it hungers for infinity. Outward in that infinity lies the Garden the sixteenth-century voyagers did not find. We no longer call it the Garden. We are sophisticated men. We call it, vaguely, "progress," because that word in itself implies the endless movement of pursuit. We have abandoned the past without realizing that without the past the pursued future has no meaning, that it leads, as Morris has anticipated, to the world of artless, dehumanized man.

III

SOME time ago there was encountered, in the litter of a vacant lot in a small American town, a fallen sign. This sign was intended to commemorate the names of local heroes who had fallen in the Second World War. But that war was over, and another had come in Korea. Probably the population of that entire town had turned over in the meantime. Tom and Joe and Isaac were events of the past, and the past of the modern world is short. The names of yesterday's heroes lay with yesterday's torn newspaper. They had served their purpose and were now forgotten.

This incident may serve to reveal the nature of what has happened, or seems to be happening, to our culture, to that world which science was to beautify and embellish. I do not say that science is responsible except in the sense that men are responsible, but men increasingly are the victims of what they themselves have created. To the student of human culture, the rise of science and its dominating role in our society presents a unique phenomenon.

Nothing like it occurs in antiquity, for in antiquity nature represented the divine. It was an object of worship. It contained mysteries. It was the mother. Today the phrase has disappeared. It is nature we shape, nature, without the softening application of the word mother, which under our control and guidance hurls the missile on its path. There has been no age in history like this one, and men are increasingly brushed aside who speak of the possibility of another road into the future.

Some time ago, in a magazine of considerable circulation, I spoke about the role of love in human society, and about pressing human problems which I felt, rightly or wrongly, would not be solved by the penetration of space. The response amazed me, in some instances, by its virulence. I was denounced for interfering with the colonization of other planets, and for corruption of the young. Most pathetically of all, several people wrote me letters in which they tried to prove, largely for their own satisfaction, that love did not exist, that parents abused and murdered their children and children their parents. They concentrated upon sparse incidents of pathological violence, and averted their eyes from the normal.

It was all too plain that these individuals were seeking rationalizations behind which they might hide from their own responsibilities. They were in the whirlpool, that much was evident. But so are we all. In 1914 the London *Times* editorialized confidently that no civilized nation would bomb open cities from the air. Today there is not a civilized nation on the face of the globe that does not take this aspect of warfare for granted. Technology demands it. In Kierkegaard's deadly future man strives, or rather ceases to strive, against himself.

But crime, moral deficiencies, inadequate ethical

standards, we are prone to accept as part of the life of man. Why, in this respect, should we be regarded as unique? True, we have had Buchenwald and the Arctic slave camps, but the Romans had their circuses. It is just here, however, that the uniqueness enters in. After the passage of three hundred years from Bacon and his followers— three hundred years on the road to the earthly Paradise—there is a rising poison in the air. It crosses frontiers and follows the winds across the planet. It is man-made; no treaty of the powers has yet halted it.

Yet it is only a symbol, a token of that vast maelstrom which has caught up states and stone-age peoples equally with the modern world. It is the technological revolution, and it has brought three things to man which it has been impossible for him to do to himself previously.

First, it has brought a social environment altering so rapidly with technological change that personal adjustments to it are frequently not viable. The individual either becomes anxious and confused or, what is worse, develops a superficial philosophy intended to carry him over the surface of life with the least possible expenditure of himself. Never before in history has it been literally possible to have been born in one age and to die in another. Many of us are now living in an age quite different from the one into which we were born. The experience is not confined to a ride in a buggy, followed in later years by a ride in a Cadillac. Of far greater significance are the social patterns and ethical adjustments which have followed fast upon the alterations in living habits introduced by machines.

Second, much of man's attention is directed exteriorly upon the machines which now occupy most of his waking hours. He has less time alone than any man before him. In dictator-controlled countries he is harangued and stirred by propaganda projected upon him by machines to which he is forced to listen. In America he sits quiescent before the flickering screen in the living room while horsemen gallop across an American wilderness long vanished in the past. In the presence of so compelling an instrument, there is little opportunity in the evenings to explore his own thoughts or to participate in family living in the way that the man from the early part of the century remembers. For too many men, the exterior world with its mass-produced daydreams has become the conqueror. Where are the eager listeners who used to throng the lecture halls; where are the workingmen's intellectual clubs? This world has vanished into the whirlpool.

Third, this outward projection of attention,

along with the rise of a science whose powers and creations seem awe-inspiringly remote, as if above both man and nature, has come dangerously close to bringing into existence a type of man who is not human. He no longer thinks in the old terms; he has ceased to have a conscience. He is an instrument of power. Because his mind is directed outward upon this power torn from nature, he does not realize that the moment such power is brought into the human domain it partakes of human freedom. It is no longer safely *within* nature; it has become violent, sharing in human ambivalence and moral uncertainty.

At the same time that this has occurred, the scientific worker has frequently denied personal responsibility for the way his discoveries are used. The scientist points to the evils of the statesmen's use of power. The statesmen shrug and remind the scientist that they are encumbered with monstrous forces that science has unleashed upon a totally unprepared public. But there are few men on either side of the Iron Curtain able to believe themselves in any sense personally responsible for this situation. Individual conscience lies too close to home, and is archaic. It is better, we subconsciously tell ourselves, to speak of inevitable forces beyond human control. When we reason thus, however, we lend powers to the whirlpool; we bring nearer the future which Kierkegaard saw, not as the *necessary* future, but one just as inevitable as man has made it.

IV

WE HAVE now glimpsed, however briefly and inadequately, the fact that modern man is being swept along in a stream of things, giving rise to other things, at such a pace that no substantial ethic, no inward stability, has been achieved. Such stability as survives, such human courtesies as remain, are the remnants of an older Christian order. Daily they are attenuated. In the name of mass man, in the name of unionism, for example, we have seen violence done and rudeness justified. I will not argue the justice or injustice of particular strikes. I can only remark that the violence to which I refer has been the stupid, meaningless violence of the rootless, nonhuman members of the age that is close to us. It is the asphalt man who defiantly votes the convicted labor boss back into office and who says: "He gets me a bigger pay check. What do I care what he does?" This is a growing aspect of modern society that runs from teen-age gangs to the corporation boards of amusement industries that deliberately plan the further debauchment of public taste.

v. *How Human Is Man?*

It is, unfortunately, the "ethic" of groups, not of society. It cannot replace personal ethic or a sense of personal responsibility for society at large. It is, in reality, group selfishness, not ethics. In the words of Max Picard, "Spirit has been divided, fragmented; here is a spirit belonging to this and to that sociological group, each group having its own peculiar little spirit, exactly what one needs in the Flight, where, in order to flee more easily, one breaks the whole up into parts; and as always happens when one separates the part from the whole . . . one magnifies the tiny part, making it ridiculously important, so that no one may notice that the tiny part is not the whole."

All over the world this fragmentation is taking place. Small nationalisms, as in Cyprus or Algiers, murder in the name of freedom. In America, child gangs battle in the streets. The group ethic as distinct from personal ethic is faceless and obscure. It is whatever its leaders choose it to mean; it destroys the innocent and justifies the act in terms of the future. In Russia this has been done on a colossal scale. The future is no more than the running of the whirlpool. It is not divinely ordained. It has been wrought by man in ignorance and folly. That folly has two faces; one is our secularized conception of progress; the other is the mass loss of personal ethic as distinguished from group ethic.

It would be idle to deny that progress has its

root in the Christian ethic, or that history, viewed as progression toward a goal, as unique rather than cyclic, is also a product of Christian thinking. There is a sense in which one can say that man entered into history through Christianity, for as Berdyaev somewhere observes, it is this religion, par excellence, which took God out of nature and elevated man above nature. The struggle for the realization of the human soul, the attempt to lift it beyond its base origins, became, in the earlier Christian centuries, the major preoccupation of the Church.

When science developed, in the hands of Bacon and his followers, the struggle for progress ceased to be an interior struggle directed toward the good life in the soul of the individual. Instead, the enormous success of the experimental method focused attention upon the power which man could exert over nature. Now he found, through Bacon's road back to the Garden, that he could share once more in that fruit of the legendary tree. With the rise of industrial science, "progress" became the watchword of the age, but it was a secularized progress. It was the increasing whirlpool of goods, cannon, bodies and yielded-up souls that an outward concentration upon the mastery of material nature was sure to bring.

Let us admit at once that the interpretation of secular progress is two-sided. If this were not so, men would more easily recognize their dilemma. Science has brought remedies for physical pain and disease; it has opened out the far fields of the universe. Gross superstition and petty dogmatism have withered under its glance. It has supplied us with fruits unseen in nature, and given an opportunity, has told us dramatically of the paradise that might be ours if we could struggle free of ancient prejudices that still beset us. No man can afford to ignore this aspect of science, no man can evade those haunting visions.

It is the roar of the whirlpool, nonetheless, that breaks now most constantly upon our listening ears, increasingly instructing us upon the most important aspect of progress—that which in secularizing the concept we have forgotten. Its sound marks the dangerous near-dissolution of man's second nature, custom. Ideas, heresies, run like wildfire and death over the crackling static of the air. They no longer pick their way slowly through the experience of generations. Tax burdens multiply and reach upward year by year as man pays for his engines of death and underwrites ever more wearily the cost of the "progress" to which this road has led him. There is no retreat. The great green forest that once surrounded us Americans and behind which we could seek refuge has been consumed. And thus, though more symbolically, has it been everywhere for man. We have re-entered nature, not like a Greek shepherd on a hillside hearing joyfully the returning pipes of Pan, but rather as an evil and precocious animal who slinks home in the night with a few stolen powers. The serenity of the gods is not disturbed. They know well on whose head the final lightning will fall.

Progress secularized, progress which pursues only the next invention, progress which pulls thought out of the mind and replaces it with idle slogans, is not progress at all. It is a beckoning mirage in a desert over which stagger the generations of men. Because man, each individual man among us, possesses his own soul and by that light must live or perish, there is no way by which Utopias—or the lost Garden itself—can be brought out of the future and presented to man. Neither can he go forward to such a destiny. Since in the world of time every man lives but one life, it is in himself that he must search for the secret of the Garden. With the fading of religious emphasis and the growth of the torrent, modern man is confused. The tumult without has obscured those voices that still cry desperately to man from somewhere within his consciousness.

V

ONE hundred years ago last autumn, Charles Darwin published the *Origin of Species*. Epic of science though it is, it was a great blow to man. Earlier, man had seen his world displaced from the center of space; he had seen the Empyrean heaven vanish to be replaced by a void filled only with the wandering dust of worlds; he had seen earthly time lengthen until man's duration within it was only a small whisper on the sidereal clock. Finally, now, he was to be taught that his trail ran backward until in some lost era it faded into the night-world of the beast. Because it is easier to look backward than to look forward, since the past is written in the rocks, this observation, too, was added to the whirlpool.

"I am an animal," man considered. It was a fair judgment, an outside judgment. Man went into the torrent along with the steel of the first ironclads and a new slogan, "the survival of the fittest." There would be one more human retreat when, in the twentieth century, human values themselves would fall under scrutiny and be judged relative, shifting and uncertain. It is the way of the torrent—everything touched by it begins to circle without direction. All is relative, there is nothing fixed, and of guilt there can, of course, remain but little. Moral responsibility has difficulty in existing consistently beside the new scientific determinism.

I remarked at the beginning of this discussion that the "within," man's subjective nature, and the things that come to him from without often bear a striking relationship. Man cannot be long studied as an object without his cleverly altering his inner defenses. He thus becomes a very difficult creature with which to deal. Let me illustrate this concretely.

Not long ago, hoping to find relief from the duties of my office, I sought refuge with my books on a campus bench. In a little while there sidled up to me a red-faced derelict whose parboiled features spoke eloquently of his particular weakness. I was resolved to resist all blandishments for a handout.

"Mac," he said, "I'm out of a job. I need help."

I remained stonily indifferent.

"Sir," he repeated.

"Uh huh," I said, unwillingly looking up from my book.

"I'm an alcoholic."

"Oh," I said. There didn't seem to be anything else to say. You can't berate a man for what he's already confessed to you.

"I'm an alcoholic," he repeated. "You know what that means? I'm a sick man. Not giving me alcohol is ill-treating a sick man. I'm a sick man. I'm an alcoholic. I have to have a drink. I'm telling you honest. It's a disease. I'm an alcoholic. I can't help myself."

"Okay," I said, "you're an alcoholic." Grudgingly I contributed a quarter to his disease and his increasing degradation. But the words stayed in my head. "I can't help myself." Let us face it. In one disastrous sense, he was probably right. At least at this point. But where had the point been reached, and when had he developed this clever neo-modern, post-Freudian panhandling lingo? From what judicious purloining of psychiatric or social-work literature and lectures had come these useful phrases?

And he had chosen his subject well. At a glance he had seen from my book and spectacles that I was susceptible to this approach. I was immersed in the modern dilemma. I could have listened, gazing into his mottled face, without an emotion if he had spoken of home and mother. But he was an alcoholic. He knew it and he guessed that I might be a scientist. He had to be helped from the outside. It was not a moral problem. He was ill.

I settled uncomfortably into my book once more, but the phrase stayed with me, "I can't help myself." The clever reversal. The outside judgment turned back and put to dubious, unethical use by the man inside.

"I can't help myself." It is the final exteriorization of man's moral predicament, of his loss of authority over himself. It is the phrase that, above all others, tortures the social scientist. In it is truth, but in it also is a dreadful, contrived folly. It is society, a genuinely sick society, saying to its social scientists, as it says to its engineers and doctors: "Help me. I'm rotten with hate and ignorance that I won't give up, but you are the doctor; fix me." This, says society, is *our* duty. We are social scientists. Individuals, poor blighted specimens, cannot assume such responsibilities. "True, true," we mutter as we read the case histories. "Life is dreadful, and yet—"

Man on the inside is quick to accept scientific judgments and make use of them. He is conditioned to do this. This new judgment is an easy one; it deadens man's concern for himself. It makes the way into the whirlpool easier. In spite of our boasted vigor we wait for the next age to be brought to us by Madison Avenue and General Motors. We do not prepare to go there by means of the good inner life. We wait, and in the meantime it slowly becomes easier to mistake longer cars or brighter lights for progress. And yet—

Forty thousand years ago in the bleak uplands of southwestern Asia, a man, a Neanderthal man, once labeled by the Darwinian proponents of struggle as a ferocious ancestral beast—a man whose face might cause you some slight uneasiness if he sat beside you—a man of this sort existed with a fearful body handicap in that ice-age world. He had lost an arm. But still he lived and was cared for. Somebody, some group of human things, in a hard, violent and stony world, loved this maimed creature enough to cherish him.

And looking so, across the centuries and the millennia, toward the animal men of the past, one can see a faint light, like a patch of sunlight moving over the dark shadows on a forest floor. It shifts and widens, it winks out, it comes again, but it persists. It is the human spirit, the human soul, however transient, however faulty men may claim it to be. In its coming man had no part. It merely came, that curious light, and man, the animal, sought to be something that no animal had been before. Cruel he might be, vengeful he might be, but there had entered into his nature a curious wistful gentleness and courage. It seemed to have little to do with survival, for such men died over and over. They did not value life compared to what they saw in themselves—that strange inner light which has come from no man knows where, and which was not made by us. It has followed us all the way from the age of ice, from the dark borders of the ancient forest into which our footprints vanish. It is in this that Kierkegaard glimpsed the eternal, the way of the heart, the way of love which is not of today, but is of the whole journey and may lead us at last to the end. Through this, he thought, the future may be conquered. Certainly it is true. For man may grow until he towers to the skies, but without this light he is nothing, and his place is nothing. Even as we try to deny the light, we know that it has made us, and what we are without it remains meaningless.

We have come a long road up from the darkness, and it well may be—so brief, even so, is the human story—that viewed in the light of history, we are still uncouth barbarians. We are potential love animals, wrenching and floundering in our larval envelopes, trying to fling off the bestial past. Like children or savages, we have delighted ourselves with technics. We have thought they alone might free us. As I remarked before, once launched on this road, there is no retreat. The whirlpool can be conquered, but only by placing it in proper perspective. As it grows, we must learn to cultivate that which must never be permitted to enter the maelstrom—ourselves. We must never accept utility as the sole reason for education. If all knowledge is of the outside, if none is turned inward, if self-awareness fades into the blind acquiescence of the mass man, then the personal responsibility by which democracy lives will fade also.

Schoolrooms are not and should not be the place where man learns only scientific techniques. They are the place where selfhood, what has been called "the supreme instrument of knowledge," is created. Only such deep inner knowledge truly expands horizons and makes use of technology, not for power, but for human happiness. As the capacity for self-awareness is intensified, so will return that sense of personal responsibility which has been well-nigh lost in the eager yearning for aggrandizement of the asphalt man. The group may abstractly desire an ethic, theologians may preach an ethic, but no group ethic ever could, or should, replace the personal ethic of individual, responsible men. Yet it is just this which the Marxist countries are seeking to destroy; and we, in a vague, good-natured indifference, are furthering its destruction by our concentration upon material enjoyment and our expressed contempt for the man who thinks, to our mind, unnecessarily.

Let it be admitted that the world's problems are many and wearing, and that the whirlpool runs fast. If we are to build a stable cultural structure above that which threatens to engulf us by changing our lives more rapidly than we can adjust our habits, it will only be by flinging over the torrent a structure as taut and flexible as a spider's web, a human society deeply self-conscious and undeceived by the waters that race beneath it, a society more literate, more appreciative of human worth than any society that has previously existed. That is the sole prescription, not for survival—which is meaningless—but for a society worthy to survive. It should be, in the end, a society more interested in the cultivation of noble minds than in change.

There is a story about one of our great atomic physicists—a story for whose authenticity I cannot vouch, and therefore I will not mention his name. I hope, however, with all my heart that it is true. If it is not, then it ought to be, for it illustrates well what I mean by a growing self-awareness, a sense of responsibility about the universe.

This man, one of the chief architects of the atomic bomb, so the story runs, was out wandering in the woods one day with a friend when he came upon a small tortoise. Overcome with pleasurable excitement, he took up the tortoise and started home, thinking to surprise his children with it. After a few steps he paused and surveyed the tortoise doubtfully.

"What's the matter?" asked his friend.

Without responding, the great scientist slowly retraced his steps as precisely as possible, and gently set the turtle down upon the exact spot from which he had taken him up.

Then he turned solemnly to his friend. "It just struck me," he said, "that perhaps, for one man, I have tampered enough with the universe." He turned, and left the turtle to wander on its way.

The man who made that remark was one of the best of the modern men, and what he had devised had gone down into the whirlpool. "I have tampered enough," he said. It was not a denial of science. It was a final recognition that science is not enough for man. It is not the road back to the waiting Garden, for that road lies through the heart of man. Only when man has recognized this fact will science become what it was for Bacon, something to speak of as "touching upon Hope." Only then will man be truly human.

VI

HOW NATURAL IS "NATURAL"?

The very design of imagination is to domesticate us in another, a celestial nature.

RALPH WALDO EMERSON

I N THE more obscure scientific circles which I frequent there is a legend circulating about a late distinguished scientist who, in his declining years, persisted in wearing enormous padded boots much too large for him. He had developed, it seems, what to his fellows was a wholly irrational fear of falling through the interstices of that largely empty molecular space which common men in their folly speak of as the world. A stroll across his living-room floor had become, for him, something as dizzily horrendous as the activities of a window washer on the Empire State Building. Indeed, with equal reason he could have passed a ghostly hand through his own ribs.

The quivering network of his nerves, the awe-inspiring movement of his thought had become a vague cloud of electrons interspersed with the light-year distances that obtain between us and the farther galaxies. This was the natural world which he had helped to create, and in which, at last, he had found himself a lonely and imprisoned occupant. All around him the ignorant rushed on their way over the illusion of substantial floors, leaping, though they did not see it, from particle to particle, over a bottomless abyss. There was even a question as to the reality of the particles which bore them up. It did not, however, keep insubstantial newspa-

pers from being sold, or insubstantial love from being made.

Not long ago I became aware of another world
perhaps equally natural and real, which man is beginning to forget. My thinking began in New England under a boat dock. The lake I speak of has
been pre-empted and civilized by man. All day
long in the vacation season high-speed motorboats,
driven with the reckless abandon common to the
young Apollos of our society, speed back and forth,
carrying loads of equally attractive girls. The
shores echo to the roar of powerful motors and the
delighted screams of young Americans with un-
counted horsepower surging under their hands.
In truth, as I sat there under the boat dock, I had
some desire to swim or to canoe in the older ways
of the great forest which once lay about this region.
Either notion would have been folly. I would have
been gaily chopped to ribbons by teen-age young-
sters whose eyes were always immutably fixed on
the far horizons of space, or upon the dials which
indicated the speed of their passing. There was another world, I was to discover, along the lake shal-
lows and under the boat dock, where the motors
could not come.

As I sat there one sunny morning when the water
was peculiarly translucent, I saw a dark shadow
moving swiftly over the bottom. It was the first sign
of life I had seen in this lake, whose shores seemed

to yield little but washed-in beer cans. By and by the gliding shadow ceased to scurry from stone to stone over the bottom. Unexpectedly, it headed almost directly for me. A furry nose with gray whiskers broke the surface. Below the whiskers green water foliage trailed out in an inverted V as long as his body. A muskrat still lived in the lake. He was bringing in his breakfast.

I sat very still in the strips of sunlight under the pier. To my surprise the muskrat came almost to my feet with his little breakfast of greens. He was young, and it rapidly became obvious to me that he was laboring under an illusion of his own, and that he thought animals and men were still living in the Garden of Eden. He gave me a friendly glance from time to time as he nibbled his greens. Once, even, he went out into the lake again and returned to my feet with more greens. He had not, it seemed, heard very much about men. I shuddered. Only the evening before I had heard a man describe with triumphant enthusiasm how he had killed a rat in the garden because the creature had dared to nibble his petunias. He had even showed me the murder weapon, a sharp-edged brick.

On this pleasant shore a war existed and would go on until nothing remained but man. Yet this creature with the gray, appealing face wanted very little: a strip of shore to coast up and down, sunlight and moonlight, some weeds from the deep water. He was an edge-of-the-world dweller, caught between a vanishing forest and a deep lake pre-empted by unpredictable machines full of chopping blades. He eyed me nearsightedly, a green leaf poised in his mouth. Plainly he had come with some poorly instructed memory about the lion and the lamb.

"You had better run away now," I said softly, making no movement in the shafts of light. "You are in the wrong universe and must not make this mistake again. I am really a very terrible and cunning beast. I can throw stones." With this I dropped a little pebble at his feet.

He looked at me half blindly, with eyes much better adjusted to the wavering shadows of his lake bottom than to sight in the open air. He made almost as if to take the pebble up into his forepaws. Then a thought seemed to cross his mind—a thought perhaps telepathically received, as Freud once hinted, in the dark world below and before man, a whisper of ancient disaster heard in the depths of a burrow. Perhaps after all this was not Eden. His nose twitched carefully; he edged toward the water.

As he vanished in an oncoming wave, there went with him a natural world, distinct from the world of girls and motorboats, distinct from the world of the professor holding to reality by some great snow-shoe effort in his study. My muskrat's shore-line universe was edged with the dark wall of hills on one side and the waspish drone of motors farther out, but it was a world of sunlight he had taken down into the water weeds. It hovered there, waiting for my disappearance. I walked away, obscurely pleased that darkness had not gained on life by any act of mine. In so many worlds, I thought, how natural is "natural"—and is there anything we can call a natural world at all?

II

Nature, contended John Donne in the seventeenth century, is the common law by which God governs us. Donne was already aware of the new science and impressed by glimpses of those vast abstractions which man was beginning to build across the gulfs of his ignorance. Donne makes, however, a reservation which rings strangely in the modern ear. If nature is the common law, he said, then Miracle is God's Prerogative.

By the nineteenth century, this spider web of common law had been flung across the deeps of space and time. "In astronomy," meditates Emerson, "vast distance, but we never go into a foreign system. In geology, vast duration, but we are never strangers. Our metaphysic should be able to follow the flying force through all its transformations."

Now admittedly there is a way in which all these

worlds are real and sufficiently natural. We can say, if we like, that the muskrat's world is naïve and limited, a fraction, a bare fraction, of the world of life: a view from a little pile of wet stones on a nameless shore. The view of the motor speedsters in essence is similar and no less naïve. All would give way to the priority of that desperate professor, striving like a tired swimmer to hold himself aloft against the soft and fluid nothingness beneath his feet. In terms of the modern temper, the physicist has penetrated the deepest into life. He has come to that place of whirling sparks which are themselves phantoms. He is close upon the void where science ends and the manifestation of God's Prerogative begins. "He can be no creature," argued Donne, "who is present at the first creation."

Yet there is a way in which the intelligence of man in this era of science and the machine can be viewed as having taken the wrong turning. There is a dislocation of our vision which is, perhaps, the product of the kind of creatures we are, or at least conceive ourselves to be. As we mentioned earlier, man, as a two-handed manipulator of the world about him, has projected himself outward upon his surroundings in a way impossible to other creatures. He has done this since the first half-human man-ape hefted a stone in his hand. He has always sought mastery over the materials of his environment, and in our day he has pierced so deeply through the screen of appearances that the age-old distinctions between matter and energy have been dimmed to the point of disappearance. The creations of his clever intellect ride in the skies and the sea's depths; he has hurled a great fragment of metal at the moon, which he once feared. He holds the heat of suns within his hands and threatens with it both the lives and happiness of his unborn descendants.

Man, in the words of one astute biologist, is "caught in a physiological trap and faced with the problem of escaping from his own ingenuity." Pascal, with intuitive sensitivity, saw this at the very dawn of the modern era in science. "There is nothing which we cannot make natural," he wrote, and then, prophetically, comes the full weight of his judgment upon man, "there is nothing natural which we do not destroy." *Homo faber*, the tool-maker, is not enough. There must be another road and another kind of man lurking in the mind of this odd creature, but whether the attraction of that path is as strong as the age-old primate addiction to taking things apart remains to be seen.

We who are engaged in the life of thought are likely to assume that the key to an understanding of the world is knowledge, both of the past and of the future—that if we had that knowledge we

would also have wisdom. It is not my intention
here to decry learning. It is only to say that we
must come to understand that learning is endless
and that nowhere does it lead us behind the exist-
ent world. It may reduce the prejudices of ignorance, set our bones, build our cities. In itself it
will never make us ethical men. Yet because ours,
we conceive, is an age of progress, and because we
know more about time and history than any men
before us, we fallaciously equate ethical advance
with scientific progress in a point-to-point relationship. Thus as society improves physically, we assume the improvement of the individual, and are
all the more horrified at those mass movements of
terror which have so typified the first half of this
century.

On the morning of which I want to speak, I was
surfeited with the smell of mortality and tired of
the years I had spent in archaeological dustbins.
I rode out of a camp and across a mountain. I
would never have believed, before it happened,
that one could ride into the past on horseback.
It is true I rode with a purpose, but that purpose
was to settle an argument within myself.

It was time, I thought, to face up to what was in
my mind—to the dust and the broken teeth and
the spilled chemicals of life seeping away into the
sand. It was time I admitted that life was of the
earth, earthy, and could be turned into a piece of

wretched tar. It was time I consented to the prop-
osition that man had as little to do with his fate
as a seed blown against a grating. It was time I
looked upon the world without spectacles and saw
love and pride and beauty dissolve into effervescing
juices. I could be an empiricist with the best of
them. I would be deceived by no more music. I
had entered a black cloud of merciless thought,
but the horse, as it chanced, worked his own way
over that mountain.

I could hear the sudden ring of his hooves as we
came cautiously treading over a tilted table of
granite, past the winds that blow on the high places
of the world. There were stones there so polished
that they shone from the long ages that the storms
had rushed across them. We crossed the divide
then, picking our way in places scoured by ancient
ice action, through boulder fields where nothing
moved, and yet where one could feel time like an
enemy hidden behind each stone.

If there was life on those heights, it was the thin
life of mountain spiders who caught nothing in
their webs, or of small gray birds that slipped
soundlessly among the stones. The wind in the
pass caught me head on and blew whatever
thoughts I had into a raveling stream behind me,
until they were all gone and there was only myself
and the horse, moving in an eternal dangerous
present, free of the encumbrances of the past.

We crossed a wind that smelled of ice from still higher snowfields, we cantered with a breeze that came from somewhere among cedars, we passed a gust like Hell's breath that had risen straight up from the desert floor. They were winds and they did not stay with us. Presently we descended out of their domain, and it was curious to see, as we dropped farther through gloomy woods and canyons, how the cleansed and scoured mind I had brought over the mountain began, like the water in those rumbling gorges, to talk in a variety of voices, to debate, to argue, to push at stones, or curve subtly around obstacles. Sometimes I wonder whether we are only endlessly repeating in our heads an argument that is going on in the world's foundations among crashing stones and recalcitrant roots.

"Fall, fall, fall," cried the roaring water and the grinding pebbles in the torrent. "Let go, come with us, come home to the place without light." But the roots clung and climbed and the trees pushed up, impeding the water, and forests filled even the wind with their sighing and grasped after the sun. It is so in the mind. One can hear the rattle of falling stones in the night, and the thoughts like trees holding their place. Sometimes one can shut the noise away by turning over on the other ear, sometimes the sounds are as dreadful as a storm in the mountains, and one lies awake, holding, like the roots that wait for daylight. It was after such a night that I came over the mountain, but it was the descent on the other side that suddenly struck me as a journey into the eons of the past.

I came down across stones dotted with pink and gray lichens—a barren land dreaming life's last dreams in the thin air of a cold and future world.

I passed a meadow and a meadow mouse in a little shower of petals struck from mountain flowers. I dismissed it—it was almost my own time—a pleasant golden hour in the age of mammals, lost before the human coming. I rode heavily toward an old age far backward in the reptilian dark.

I was below timber line and sinking deeper and deeper into the pine woods, whose fallen needles lay thick and springy over the ungrassed slopes. The brown needles and the fallen cones, the stiff, endless green forests were a mark that placed me in the Age of Dinosaurs. I moved in silence now, waiting a sign. I saw it finally, a green lizard on a stone. We were far back, far back. He bobbed his head uncertainly at me, and I reined in with the nostalgic intent, for a moment, to call him father, but I saw soon enough that I was a ghost who troubled him and that he would wish me, though he had not the voice to speak, to ride on. A man who comes down the road of time should not expect to converse—even with his own kin. I made a brief, uncertain sign of recognition, to which he did not respond, and passed by. Things grew more lonely. I was coming out upon the barren ridges of an old sea beach that rose along the desert floor. Life was small and grubby now. The hot, warning scarlet of peculiar desert ants occasionally flashed among the stones. I had lost all trace of myself and thought regretfully of the lizard who might have directed me.

A turned-up stone yielded only a scorpion who curled his tail in a kind of evil malice. I surveyed him reproachfully. He was old enough to know the secret of my origin, but once more an ancient, bitter animus drawn from that poisoned soil possessed him and he raised his tail. I turned away. An enormous emptiness by degrees possessed me. I was back almost, in a different way, to the thin air over the mountain, to the end of all things in the cold starlight of space.

I passed some indefinable bones and shells in the salt-crusted wall of a dry arroyo. As I reined up, only sand dunes rose like waves before me and if life was there it was no longer visible. It was like coming down to the end—to the place of fires where we began. I turned about then and let my gaze go up, tier after tier, height after height, from crawling desert bush to towering pine on the great slopes far above me.

In the same way animal life had gone up that road from these dry, envenomed things to the deer nuzzling a fawn in the meadows far above. I had come down the whole way into a place where one could lift sand and ask in a hollow, dust-shrouded whisper, "Life, what is it? Why am I here? Why am I here?"

And my mind went up that figurative ladder of the ages, bone by bone, skull by skull, seeking an answer. There was none, except that in all that downrush of wild energy that I had passed in the canyons there was this other strange organized stream that marched upward, gaining a foothold here, tossing there a pine cone a little farther upward into a crevice in the rock.

And again one asked, not of the past this time, but of the future, there where the winds howled through open space and the last lichens clung to the naked rock, "Why did we live?" There was no answer I could hear. The living river flowed out of nowhere into nothing. No one knew its source or its departing. It was an apparition. If one did not see it there was no way to prove that it was real.

No way, that is, except within the mind itself. And the mind, in some strange manner so involved with time, moving against the cutting edge of it like the wind I had faced on the mountain, has yet its own small skull-borne image of eternity. It is not alone that I can reach out and receive within my head a handsbreadth replica of the far fields of the universe. It is not because I can touch a trilobite and know the fall of light in ages before my birth. Rather, it lies in the fact that the human mind can transcend time, even though trapped, to all appearances, within that medium. As from some remote place, I see myself as child and young man, watch with a certain dispassionate objectivity the violence and tears of a remote youth who was once I, shaping his character, for good or ill, toward the creature he is today. Shrinking, I see him teeter at the edge of abysses he never saw. With pain I acknowledge acts undone that might have saved and led him into some serene and noble pathway. I move about him like a ghost, that vanished youth. I exhort, I plead. He does not hear me. Indeed, he too is already a ghost. He has become me. I am what I am. Yet the point is, we are not wholly given over to time—if we were, such acts, such leaps through that gray medium, would be impossible. Perhaps God himself may rove in similar pain up the dark roads of his universe. Only how would it be, I wonder, to contain at once both the beginning and the end, and to hear, in helplessness perhaps, the fall of worlds in the night?

This is what the mind of man is just beginning to achieve—a little microcosm, a replica of whatever it is that, from some unimaginable "outside," contains the universe and all the fractured bits of seeing which the world's creatures see. It is not necessary to ride over a mountain range to experience historical infinity. It can descend upon one in the lecture room.

I find it is really in daylight that the sensation I am about to describe is apt to come most clearly upon me, and for some reason I associate it extensively with crowds. It is not, you understand, an hallucination. It is a reality. It is, I can only say with difficulty, a chink torn in a dimension life was never intended to look through. It connotes a sense beyond the eye, though the twenty years' impressions are visual. Man, it is said, is a time-binding animal, but he was never intended for this. Here is the way it comes.

I mount the lecturer's rostrum to address a class. Like any work-worn professor fond of his subject, I fumble among my skulls and papers, shuffle to the blackboard and back again, begin the patient translation of three billion years of time into chalk scrawls and uncertain words ventured timidly to a sea of young, impatient faces. Time does not frighten them, I think enviously. They have, most of them, never lain awake and grasped the sides of a cot, staring upward into the dark while the slow clock strokes begin.

"Doctor." A voice diverts me. I stare out nearsightedly over the class. A hand from the back row gesticulates. "Doctor, do you believe there is a direction to evolution? Do you believe, Doctor . . . Doctor, do you believe? . . ." Instead of the words, I hear a faint piping, and see an eager scholar's face squeezed and dissolving on the body of a chest-thumping ape. "Doctor, is there a direction?"

I see it then—the trunk that stretches monstrously behind him. It winds out of the door, down dark and obscure corridors to the cellar, and vanishes into the floor. It writhes, it crawls, it barks and snuffles and roars, and the odor of the swamp exhales from it. That pale young scholar's face is the last bloom on a curious animal extrusion through time. And who among us, under the cold persuasion of the archaeological eye, can perceive which of his many shapes is real, or if, perhaps, the entire shape in time is not a greater and more curious animal than its single appearance?

I too am aware of the trunk that stretches loathsomely back of me along the floor. I too am a many-visaged thing that has climbed upward out of the dark of endless leaf falls, and has slunk, furred, through the glitter of blue glacial nights. I, the professor, trembling absurdly on the platform with my book and spectacles, am the single philosophical animal. I am the unfolding worm, and mud fish, the weird tree of Igdrasil shaping itself endlessly out of darkness toward the light.

I have said this is not an illusion. It is when one sees in this manner, or a sense of strangeness halts one on a busy street to verify the appearance of one's fellows, that one knows a terrible new sense has opened a faint crack into the Absolute. It is in this way alone that one comes to grips with a great mystery, that life and time bear some curious relationship to each other that is not shared by inanimate things.

It is in the brain that this world opens. To our descendants it may become a commonplace, but me, and others like me, it has made a castaway. I have no refuge in time, as others do who troop homeward at nightfall. As a result, I am one of those who linger furtively over coffee in the kitchen at bedtime or haunt the all-night restaurants. Nevertheless, I shall say without regret: there are hazards in all professions.

I I I

IT MAY seem at this point that I have gone considerably round about in my examination of the natural world. I have done so in the attempt to indicate that the spider web of law which has been flung, as Emerson indicated, across the deeps of time and space and between each member of the living world, has brought us some quite remarkable, but at the same time disquieting, knowledge. In rapid summary, man has passed from a natural world of appearances invisibly controlled by the caprice of spirits to an astronomical universe visualized by Newton, through the law of gravitation, as operating with the regularity of a clock.

Newton, who remained devout, assumed that God, at the time of the creation of the solar system, had set everything to operating in its proper orbit. He recognized, however, certain irregularities of planetary movement which, in time, would lead to a disruption of his perfect astronomical machine. It was here, as a seventeenth-century scholar, that he felt no objection to the notion that God interfered at periodic intervals to correct the deviations of the machine.

A century later Laplace had succeeded in dispensing with this last vestige of divine intervention. Hutton had similarly dealt with supernaturalism in earth-building, and Darwin, in the nineteenth century, had gone far toward producing a similar mechanistic explanation of life. The machine that began in the heavens had finally been installed in the human heart and brain. "We can make everything natural," Pascal had truly said, and surely the more naïve forms of worship of the unseen are vanishing.

Yet strangely, with the discovery of evolutionary, as opposed to purely durational, time, there emerges into this safe-and-sane mechanical universe something quite unanticipated by the eighteenth-century rationalists—a kind of emergent, if not miraculous, novelty.

I know that the word "miraculous" is regarded dubiously in scientific circles because of past quarrels with theologians. The word has been defined, however, as an event transcending the known laws of nature. Since, as we have seen, the laws of nature have a way of being altered from one generation of scientists to the next, a little taste for the miraculous in this broad sense will do us no harm. We forget that nature itself is one vast miracle transcending the reality of night and nothingness. We forget that each one of us in his personal life repeats that miracle.

Whatever may be the power behind those dancing motes to which the physicist has penetrated, it makes the light of the muskrat's world as it makes the world of the great poet. It makes, in fact, all of the innumerable and private worlds which exist in the heads of men. There is a sense in which we can say that the planet, with its strange freight of life, is always just passing from the unnatural to the natural, from that Unseen which man has always reverenced to the small reality of the day. If all life were to be swept from the world, leaving only its chemical constituents, no visitor from another star would be able to establish the reality of such a phantom. The dust would lie without visible protest, as it does now in the moon's airless craters, or in the road before our door.

Yet this is the same dust which, dead, quiescent and unmoving, when taken up in the process known as life, hears music and responds to it, weeps bitterly over time and loss, or is oppressed by the looming future that is, on any materialist terms, the veriest shadow of nothing. How natural was man, we may ask, until he came? What forces dictated that a walking ape should watch the red shift of light beyond the island universes or listen by carefully devised antennae to the pulse of unseen stars? Who, whimsically, conceived that the plot of the world should begin in a mud puddle and end—where, and with whom? Men argue learnedly over whether life is chemical chance or antichance, but they seem to forget that the life *in* chemicals may be the greatest chance of all, the most mysterious and unexplainable property in matter.

"The special value of science," a perceptive philosopher once wrote, "lies not in what it makes of the world, but in what it makes of the knower." Some years ago, while camping in a vast eroded area in the West, I came upon one of those unlikely sights which illuminate such truths.

I suppose that nothing living had moved among those great stones for centuries. They lay toppled against each other like fallen dolmens. The huge

stones were beasts, I used to think, of a kind man
ordinarily lived too fast to understand. They
seemed inanimate because the tempo of the life in
them was slow. They lived ages in one place and
moved only when man was not looking. Sometimes
at night I would hear a low rumble as one drew
itself into a new position and subsided again. Sometimes I found their tracks ground deeply into the
hillsides.

It was with considerable surprise that while trav-
ersing this barren valley I came, one afternoon,
upon what I can only describe as a very remarkable
sight. Some distance away, so far that for a little
space I could make nothing of the spectacle, my
eyes were attracted by a dun-colored object about
the size of a football, which periodically bounded
up from the desert floor. Wonderingly, I drew
closer and observed that something ropelike which
glittered in the sun appeared to be dangling from
the ball-shaped object. Whatever the object was,
it appeared to be bouncing faster and more desperately as I approached. My surroundings were
such that this hysterical dance of what at first
glance appeared to be a common stone was quite
unnerving, as though suddenly all the natural objects in the valley were about to break into a jig.
Going closer, I penetrated the mystery.

The sun was sparkling on the scales of a huge
blacksnake which was partially looped about the

body of a hen pheasant. Desperately the bird tried to rise, and just as desperately the big snake coiled and clung, though each time the bird, falling several feet, was pounding the snake's body in the gravel. I gazed at the scene in astonishment. Here in this silent waste, like an emanation from nowhere, two bitter and desperate vapors, two little whirlwinds of contending energy, were beating each other to death because their plans—something, I suspected, about whether a clutch of eggs was to turn into a thing with wings or scales—this problem, I say, of the onrushing nonexistent future, had catapulted serpent against bird.

The bird was too big for the snake to have had it in mind as prey. Most probably, he had been intent on stealing the pheasant's eggs and had been set upon and pecked. Somehow in the ensuing scuffle he had flung a loop over the bird's back and partially blocked her wings. She could not take off, and the snake would not let go. The snake was taking a heavy battering among the stones, but the high-speed metabolism and tremendous flight exertion of the mother bird were rapidly exhausting her. I stood a moment and saw the bloodshot glaze deepen in her eyes. I suppose I could have waited there to see what would happen when she could not fly; I suppose it might have been worth scientifically recording. But I could not stand that ceaseless, bloody pounding in the gravel. I thought of the eggs somewhere about, and whether they were to elongate and writhe into an armor of scales, or eventually to go whistling into the wind with their wild mother.

So I, the mammal, in my way supple, and less bound by instinct, arbitrated the matter. I unwound the serpent from the bird and let him hiss and wrap his battered coils around my arm. The bird, her wings flung out, rocked on her legs and gasped repeatedly. I moved away in order not to drive her further from her nest. Thus the serpent and I, two terrible and feared beings, passed quickly out of view.

Over the next ridge, where he could do no more damage, I let the snake, whose anger had subsided, slowly uncoil and slither from my arm. He flowed away into a little patch of bunch grass—aloof, forgetting, unaware of the journey he had made upon my wrist, which throbbed from his expert constriction. The bird had contended for birds against the oncoming future; the serpent writhing into the bunch grass had contended just as desperately for serpents. And I, the apparition in that valley—for what had I contended?—I who contained the serpent and the bird and who read the past long written in their bodies.

Slowly, as I sauntered dwarfed among overhanging pinnacles, as the great slabs which were the visible remnants of past ages laid their enormous shadows rhythmically as life and death across my face, the answer came to me. Man could contain more than himself. Among these many appearances that flew, or swam in the waters, or wavered momentarily into being, man alone possessed that unique ability.

The Renaissance thinkers were right when they said that man, the Microcosm, contains the Macrocosm. I had touched the lives of creatures other than myself and had seen their shapes waver and blow like smoke through the corridors of time. I had watched, with sudden concentrated attention, myself, this brain, unrolling from the seed like a genie from a bottle, and casting my eyes forward, I had seen it vanish again into the formless alchemies of the earth.

For what then had I contended, weighing the serpent with the bird in that wild valley? I had struggled, I am now convinced, for a greater, more comprehensive version of myself.

IV

I AM a man who has spent a great deal of his life on his knees, though not in prayer. I do not say this last pridefully, but with the feeling that the posture, if not the thought behind it, may have had some final salutary effect. I am a naturalist and a fossil hunter, and I have crawled most of the way through life. I have crawled downward into holes without a bottom, and upward, wedged into crevices where the wind and the birds scream at you until the sound of a falling pebble is enough to make the sick heart lurch. In man, I know now, there is no such thing as wisdom. I have learned this with my face against the ground. It is a very difficult thing for a man to grasp today, because of his power; yet in his brain there is really only a sort of universal marsh, spotted at intervals by quaking green islands representing the elusive stability of modern science—islands frequently gone as soon as glimpsed.

It is our custom to deny this; we are men of precision, measurement and logic; we abhor the unexplainable and reject it. This, too, is a green island. We wish our lives to be one continuous growth in knowledge; indeed, we expect them to be. Yet well over a hundred years ago Kierkegaard observed that maturity consists in the discovery that "there comes a critical moment where everything is reversed, after which the point becomes to understand more and more that there is something which cannot be understood."

When I separated the serpent from the bird and released them in that wild upland, it was not for knowledge; not for anything I had learned in science. Instead, I contained, to put it simply, the serpent and the bird. I would always contain them. I was no longer one of the contending vapors; I had embraced them in my own substance and, in some insubstantial way, reconciled them, as I had sought reconciliation with the muskrat on the shore. I had transcended feather and scale and gone beyond them into another sphere of reality. I was trying to give birth to a different self whose only expression lies again in the deeply religious words of Pascal, "You would not seek me had you not found me."

I had not known what I sought, but I was aware at last that something had found me. I no longer believed that nature was either natural or unnatural, only that nature now appears natural to man. But the nature that appears natural to man is another version of the muskrat's world under the boat dock, or the elusive sparks over which the physicist made his trembling passage. They were appearances, specialized insights, but unreal because in the constantly onrushing future they were swept away.

What had become of the natural world of that gorilla-headed little ape from which we sprang—that dim African corner with its chewed fish bones and giant ice-age pigs? It was gone more utterly than my muskrat's tiny domain, yet it had given birth to an unimaginable thing—ourselves—something overreaching the observable laws of that far epoch. Man since the beginning seems to be awaiting an event the nature of which he does not know. "With reference to the near past," Thoreau once shrewdly commented, "we all occupy the region of common sense, but in the prospect of the future we are, by instinct, transcendentalists." This is the way of the man who makes nature "natural." He stands at the point where the miraculous comes into being, and after the event he calls it "natural." The imagination of man, in its highest manifestations, stands close to the doorway of the infinite, to the world beyond the nature that we know. Perhaps, after all, in this respect man constitutes the exertion of that act which Donne three centuries ago called God's Prerogative.

Man's quest for certainty is, in the last analysis, a quest for meaning. But the meaning lies buried within himself rather than in the void he has vainly searched for portents since antiquity. Perhaps the first act in its unfolding was taken by a raw beast with a fearsome head who dreamed some difficult and unimaginable thing denied his fellows. Perhaps the flashes of beauty and insight which trouble us so deeply are no less prophetic of what the race might achieve. All that prevents us is doubt— the power to make everything natural without the accompanying gift to see, beyond the natural, to that inexpressible realm in which the words "natural" and "supernatural" cease to have meaning.

Man, at last, is face to face with himself in natural guise. "What we make natural, we destroy," said Pascal. He knew, with superlative insight, man's complete necessity to transcend the worldly image that this word connotes. It is not the outward powers of man the toolmaker that threaten us. It is a growing danger which has already afflicted vast areas of the world—the danger that we have created an unbearable last idol for our worship. That idol, that uncreate and ruined visage which confronts us daily, is no less than man made natural. Beyond this replica of ourselves, this countenance already grown so distantly inhuman that it terrifies us, still beckons the lonely figure of man's dreams. It is a nature, not of this age, but of the becoming—the light once glimpsed by a creature just over the threshold from a beast, a despairing cry from the dark shadow of a cross on Golgotha long ago.

Man is not totally compounded of the nature we profess to understand. Man is always partly of the future, and the future he possesses a power to shape. "Natural" is a magician's word—and like all such entities, it should be used sparingly lest there arise from it, as now, some unglimpsed, unintended world, some monstrous caricature called into being by the indiscreet articulation of worn syllables. Perhaps, if we are wise, we will prefer to stand like those forgotten humble creatures who poured little gifts of flints into a grave. Perhaps there may come to us then, in some such moment, a ghostly sense that an invisible doorway has been opened —a doorway which, widening out, will take man beyond the nature that he knows.

BIBLIOGRAPHY

BLYTH, EDWARD, "An Attempt to Classify the Variations of Animals, etc.," *Magazine of Natural History*, Vol. 8 (1835), pp. 40–53.

BLYTH, EDWARD, "On the Psychological Distinctions Between Man and All Other Animals, etc.," *Magazine of Natural History*, Vol. 1 N.S. (1837), Parts I, II, III.

BRÜCKNER, JOHN, *A Philosophical Survey of the Animal Creation*, London, 1768.

CANNON, H. GRAHAM, "What Lamarck Really Said," Proceedings of the Linnean Society of London, 168th Session, 1955–56, Parts I, II.

COLERIDGE, S. T., *Philosophical Lectures 1818–1819*, London, Pilot Press, 1949.

DARLINGTON, C. D., *Darwin's Place in History*, Oxford, Blackwell, 1959.

DARLINGTON, WILLIAM, *Memorials of John Bartram and Humphrey Marshall*, Philadelphia, 1849.

DART, RAYMOND and CRAIG, DENNIS, *Adventures with the Missing Link*, New York, Harper & Brothers, 1959.

DE BEER, SIR GAVIN, "Darwin's Notebooks on Transmutation of Species," Bulletin of the British Museum of Natural History, Historical Series, Vol. 2, Nos. 2, 3 (1960).

EISELEY, LOREN, "Charles Darwin, Edward Blyth and the Theory of Natural Selection," *Proceedings of the American Philosophical Society*, Vol. 103 (1959), pp. 94–158.

——— "Charles Lyell," *Scientific American*, Vol. 201 (1959), pp. 98–101.

Bibliography

FALCONER, HUGH, *Paleontological Memoirs,* Vols. I and II, London, Robert Hardwicke, 1868.

"James Hutton 1726–1797: Commemoration of the 150th Anniversary of His Death," Proceedings of the Royal Society of Edinburgh, Vol. 63 (1949), pp. 351–400.

MORRIS, WRIGHT, *The Territory Ahead,* New York, Harcourt, Brace and Company, 1958.

PICARD, MAX, *The Flight from God,* Chicago, Henry Regnery Co., 1951.

TROW-SMITH, ROBERT, *A History of British Livestock Husbandry* 1700–1900, London, Routledge and Kegan Paul, 1959.

LOREN C. EISELEY, University Professor of Anthropology and the History of Science at the University of Pennsylvania, is the author of *The Immense Journey* and *Darwin's Century. Darwin's Century* received the Athenaeum of Philadelphia Award for 1958 and the Phi Beta Kappa Award in Science for 1958. Dr. Eiseley has published essays in *The American Scholar, Scientific American, Science* and *Harper's Magazine.*

Atheneum Paperbacks

THE WORLDS OF NATURE AND MAN

LIFE SCIENCES AND ANTHROPOLOGY